Praise for *The*

"Hammack writes with admirabl[e]

—*Kirkus Reviews*

"This book unravels the mysteries behind humanity's greatest masterpieces. Bill Hammack is not just an expert on designing and building—he's a gifted communicator. His infectious enthusiasm leaps off the page as he reveals how you can learn to think like an engineer even if you don't love math."

—Adam Grant, #1 *New York Times* bestselling author
of *Think Again* and host of the TED podcast *Re:Thinking*

"Bill Hammack's book is a fascinating journey about how engineers and inventors actually work and proves that everyone—engineers, scientists, and even you and me—can create useful things by tinkering. A great book and enjoyable read."

—Don Norman, bestselling author of *The Design
of Everyday Things* and *Design for a Better World*

"A fascinating case that engineering isn't the same as science in this sweeping history... Hammack brilliantly delineates the role of trial and error in human progress and presents a knockout argument that a perfect understanding of the world is not a prerequisite to innovation. This clever and curious account delivers."

—*Publishers Weekly*, Starred Review

"A fascinating description of how engineers work, full of lively stories from engineering history, and an enlightening read."

—Roland Ennos, author of *The Age of Wood*

"This brilliant book is packed with inspiring stories about clever, determined, patient but ultimately practical people. Hammack is on a mission to reshape the accepted science narrative."

—Richard Mainwaring, author of
What the Ears Hears (and Doesn't)

"A delightful and illuminating tour of the clever solutions that engineers have devised through the centuries. This book brims with fascinating historical accounts of everything from an ancient grape juice-squeezing apparatus to a precisely-controlled 'candle clock,' while also giving us new appreciation for the humble everyday objects in our homes—bikes, microwaves, laundry detergent—that we take for granted."

—Roberta Kwok, award-winning science journalist

THE THINGS WE MAKE

The **UNKNOWN HISTORY** of **INVENTION**

from Cathedrals to Soda Cans

BILL HAMMACK, PhD

sourcebooks

This publication is designed to provide accurate and authoritative information in regard
to the subject matter covered. It is sold with the understanding that the publisher is not
engaged in rendering legal, accounting, or other professional service. If legal advice
or other expert assistance is required, the services of a competent professional person
should be sought.—*From a Declaration of Principles Jointly Adopted by a Committee
of the American Bar Association and a Committee of Publishers and Associations*

References to internet websites (URLs) were accurate at the time of
writing. Neither the author nor Sourcebooks is responsible for URLs that
may have expired or changed since the manuscript was prepared.

Published by Sourcebooks
P.O. Box 4410, Naperville, Illinois 60567-4410
(630) 961-3900
sourcebooks.com

Originally published as *The Things We Make* in 2023 in
the United States of America by Sourcebooks.

Cataloging-in-Publication Data is on file with the Library of Congress.

Printed and bound in the United States of America.
LSC 10 9 8 7 6 5 4 3 2 1

To Billy Vaughn Koen
for his pioneering book
*Discussion of the Method: Conducting the
Engineer's Approach to Problem Solving.*
Its influence graces every page of this book.

"Nearly everyone believes, falsely, that technology is applied science... Technology is more closely related to art than to science—not only materially, because art must somehow involve the selection and manipulation of matter, but conceptually as well, because the technologist, like the artist, must work with many unanalyzable complexities. Another popular misunderstanding today is the belief that technology is inherently ugly and unpleasant, whereas a moment's reflection will show that technology underlies innumerable delightful experiences as well as the greatest art, whether expressed in object, word, sound or environment."

Cyril Stanley Smith, *A Search for Structure*

CONTENTS

INTRODUCTION

When I visit Paris, my first stop is to tour the thirteenth-century cathedral of the Palais de la Cité, the Sainte-Chapelle. I enter the lower chapel just as a servant of France's thirteenth-century monarchs would have done in pious worship. Supplicants must admire for a moment its gilded arches and deep blue ceiling showcasing the wealth and power of the medieval French royalty, then climb a circular staircase, its stone walls a dungeon-like dull gray. I clutch the rail to keep balance on the uneven steps, eager to reach the upper chapel—reserved in the past for the king and queen alone but now open to all in service of the French Revolution's egalitarian values.

I pass a small sign that commands *silence* as I step into the upper chapel. When I do, I am, in the words of a fourteenth-century French philosopher, "entering into one of the best chambers of Paradise."[1] And indeed the chapel stuns all visitors into silence, never mind the sign. Once there, I settle into a pew and gaze at the chapel's four-hundred-ton stone ceiling and walls. The slenderest of pillars, slightly less than a foot in diameter, support an array of stained-glass windows that transform sunlight into a diffused red, blue, and gold glimmering on the chapel's varicolored sculptures and gilded arches. This light, a hallmark of Gothic architecture, was a sharp departure from the dark, somber buildings of the Romans in late antiquity. The immense Roman structures—the Pantheon in Rome and Hagia Sophia in Constantinople—established an earthly empire with their heavy, thick walls, while the light streaming through the translucent Gothic ceiling of the Sainte-Chapelle symbolized an otherworldly spiritual power. What I see, though, as an engineer, is an exemplar of the strategy used from the dawn of humankind to today by *all* engineers to create objects and systems.

Naively we assume the products of engineers arise from the scientific method. There's an old, slightly bitter joke among engineers about the relationship between science and engineering: "If it's a success, then it's a scientific miracle; if a disaster, then an engineering failure." This joke highlights the fact that successful technologies are invisible: we think of our furnace only when it fails. Less naively, perhaps, we assume the secret of engineering lies in the mastery of arcane realms of knowledge—sophisticated calculus and powerful computing science implemented by a dispassionate, almost mechanical person—but the power of engineers lies in their *method*, a method used long before sophisticated mathematics and computers. Indeed, modern engineers use sophisticated mathematics and a thorough knowledge of the strength of a steel beam or a slab of reinforced concrete to design a structure that withstands forces far beyond what it will ever suffer in its lifetime. The safety margin for a skyscraper requires it to hold against what engineers call a "hundred-year wind"—the strongest wind predicted to occur once every century in the skyscraper's locale. These sophisticated tools and techniques, though, are not the engineering method. The Sainte-Chapelle was designed and constructed by a team of builders who had never learned basic arithmetic or the geometry taught today in third grade. Even if they were lucky enough to be educated in reading, writing, and math, they would use them without a standardized length of the foot, as it varied from region to region. Yet medieval engineers understood stone structures so well that only a small fraction of cathedrals collapsed in their lifetime of service, before centuries of weathering and neglect after the Reformation compromised them. Their design and construction, noted a twentieth-century engineer schooled in modern construction methods, were "by almost any yardstick... almost perfect."[2] In many ways, the Sainte-Chapelle's builders had, without modern tools and techniques, a deeper understanding of stone-based architecture than today's engineers would ever hope to develop; few today would stake their reputation on building a stone cathedral (a problem for their restoration and preservation). This highlights the notion that what defines engineering, then, is not the tools—not the computer algorithms, structural analysis, or scientific knowledge of construction materials—but the method.

The purpose of this book is to lift the veil and show the engineering method in all its glory. This will reveal the creativity of engineers, demonstrate the pinnacle of the suppleness of the human mind, and lay a foundation about how to think about technology—how to decide its proper use, aid it in fulfilling its promise, but also understand its limitations. So this is not a history of engineering, although it draws from all periods of history, nor is it a celebration of "the rise of the West"—examples abound from other cultures—but a description of engineering, a deep look at its foundations, and an examination of how it can be used to shape our world.

To illustrate the method, there is no better example than the design of a cathedral. It strips bare the tools often confused for the engineering method—scientific inquiry, mathematical manipulation—to expose what lies at the heart of the method: a surprisingly simple notion called a "rule of thumb."

1
THE INVISIBLE METHOD

How to Build a Cathedral without
Mathematics, Science, or a Yardstick

A cathedral, even under construction, was a startling site of order in the chaos of a medieval city. The church's regular proportions and clean stone contrasted with the filth surrounding it. The ground around the site was a sickly mixture of mud, refuse, broken crockery, rotting meat, and human feces—the detritus from a constant hum of human activity. Thousands of people filled a medieval city: farmers guiding ponies and packhorses laden with grain or steering carts overloaded with eggs, milk, and cheese, and shepherds coaxing sheep and cattle to market. To escape this mire, children often climbed the half-built walls of the rising cathedral, undeterred by the prickly briars placed on them by the town authorities.

Like the town, the work site buzzed with activity, but here order was kept under the command of the head mason—a man whose job combined five modern roles: engineer, architect, materials contractor, building contractor, and construction supervisor.[1] These responsibilities gave him over one hundred workers to manage, who swarmed over the work site in a cacophonous chaos: the squeal of the grindstone as blacksmiths sharpened chisels, the hiss of a furnace as they forged tools, the hammer blows of plumbers as they shaped lead into eaves and gutters, the snap of a whip as a worker drove oxen to power a giant crane that lifted stone blocks, and the rasp of saws as carpenters cut boards for the intricate scaffolding that held the stone blocks as they were laid—a cathedral's construction gobbled some four thousand trees.[2]

Almost masked by this din were the intermittent sharp pings of hammers coming from a small thatch-roofed building. In this lodge attached to one of the cathedral's walls, ten or so masons worked at benches, chiseling many thousands of limestone blocks into the shapes dictated by a set of wooden templates. As they worked, the masons chattered in Hungarian, Polish, German, French, and Dutch. This itinerant workforce easily found employment—France alone constructed ten churches a year—and these elite workers were well paid. The masons took home almost twice as much as the carpenters, who themselves earned about three or four times the wages of unskilled workers, like the women who collected the moss the masons used for bedding roof tiles. The masons worked from sunrise to sunset with time off for breakfast, lunch, and a drink; in summer, they also broke for a *sieste*. Every fourteen days, their work finished two hours early so the masons could go to the baths—a perk available only to them. Unlike the carpenters and other workers, the masons all wore the same type of clothes issued by the cathedral's patron: leather hoods to protect their shoulders when carrying stones, gloves, boots for wet weather, straw hats for summer, and a robe. As these masons worked, the head mason, his robe trimmed in fur, walked from bench to bench holding in his gloved hand a *baguette*, French for *rod* or *wand*, here meaning a long, unmarked iron rod used as the standard measure for that particular cathedral. He hovered over the younger masons as they carved the stones, correcting their use of templates to shape the stones as he pointed with his baguette. His repeated refrain was "cut it there for me."[3]

This head mason, the central figure in the construction of a cathedral, commanded benefits far surpassing those of his journeyman masons: a food allowance, fodder for his horse, sometimes freedom from taxes for life, and often payment in silver, linen, wood, shoes, meat, salt, and candles as a guarantee against depreciating currency. With these fringe benefits, patrons hired a versatile worker. Head masons designed the machines that transferred stone from quarry to job site and the cranes that lifted the stones, and one master even fixed the plumbing of the king of England's toilet. But the head mason's most important task was creating the wooden templates used to guide the journeyman masons as they cut the stones. Aware of the acute focus needed, the younger masons left their master alone when the Latin phrase *in trasura* buzzed around the lodge. All then knew the master was working in the tracing room.

In the tracing room, which we would today call a drawing room, the head mason laid out thin planks of oak, fir, or pine specially imported for his use from Norway, Russia, Germany, and the Low Countries. He chalked on the boards the shapes for the faces of the stones, constructing those shapes using only a compass (dividers), straightedge, and rope. The likely illiterate head mason created these shapes without blueprints or even much of a written design. Instead, in his mind, he decomposed the construction into thousands of three-dimensional puzzle pieces that, when assembled, formed a complete cathedral. Because the building plan existed almost entirely in the head mason's memory, patrons often insisted that the master be on-site during construction, although a head mason could be in such demand that he often traveled widely throughout Europe to work on several buildings at a time. But the mason's design was not mere puzzle assembly. The result had to be a unique monument to the noble patron's wealth and piety that stood for ages. As one head mason advised his trainee son, "An honorable work glorifies its master," adding, "*if* it stands up."[4] In the shaping of the templates, the head mason maintained and transmitted the knowledge necessary to create a stable structure.

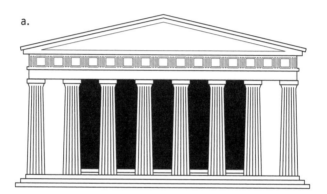

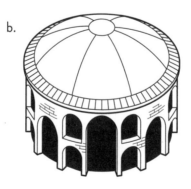

(a.) The Parthenon, an example of post-and-lintel construction. (b.) The dome of the Pantheon, which is an arch spun around its center.

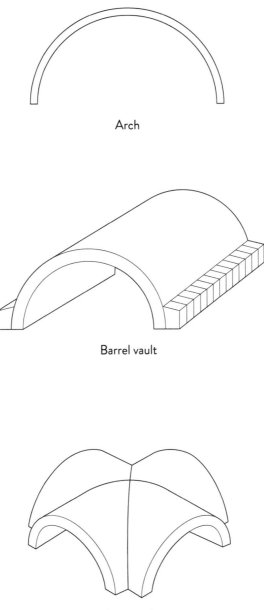

Arch

Barrel vault

Groin vault

The arch and its role in creating vaults.

To support the large interior spaces of his cathedral, a head mason could choose from two techniques that evolved as architecture moved from merely shelter, in the form of huts and tents, to grand public structures. They could use the method that the Greeks honed to perfection in the Parthenon: a post-and-lintel system, what we would call a roof supported by columns. Although the Greeks made stunning structures, this method creates buildings with small, dark interior rooms, as the columns fill the available lateral space. An architect designing for a larger spanned space would find that the slab of stone forming the roof needed more frequently placed columns to support it. For even larger buildings, then, this method becomes uneconomic and unpleasing. The post-and-lintel returned in the nineteenth century with the development of cast-iron columns and girders, improved later by high-quality steel beams, but builders intending to create grand monuments in antiquity could only turn to the option perfected by the Romans: the arch.

Roman emperors knew that the look of a building mattered. Architecture was a visual cue to the people that they lived in a Roman city, imbued with and ruled by Roman power—that they were and would be either Roman citizens or Roman subjects forever. And while the Romans liberally adopted the post-and-lintel from the Greeks, the unique imperial style was one of curves, buildings with large, centralized spaces on a huge scale covered by domes as exemplified by the Pantheon, the Temple of the Gods in Rome. Its rotunda is capped with a 144-foot diameter concrete dome, found so beautiful by Michelangelo that he called it "angelic" and "not of human design."[5] Their imperial architecture featured, on a heroic scale, the semicircular arch and its three-dimensional manifestations, the vault and the dome, which enable spaciousness and grandeur far beyond what is possible with post-and-lintel thanks to the arch's marriage of geometry and stability.

But for the masons of the Middle Ages, this type of arch presented a problem: the semicircular arch's height is always half that of its width. The head masons wanted to create churches that soared, but if they used the classical Roman arch, for every foot they added to a church's height, its width would expand by two.

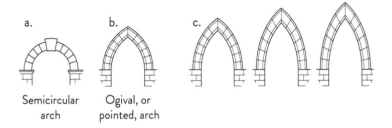

a.   b.   c.

Semicircular   Ogival, or
arch   pointed, arch

(a.) The Romans extensively used the semicircular arch. (b.) Medieval master masons favored the pointed arch, which came from the Islamic world. (c.) The pointed arch, unlike the semicircular arch, can reach any height for the same span.

To keep cathedrals from ballooning to sprawl across city blocks, they turned to an Islamic modification: the "ogival," or pointed, arch. This style, notable for its tall and slender shape, was first used in Buddhist India in the second century AD and from there spread to the near East by the seventh century, just in time for the newly created Islamic world to use in the construction of mosques. As true of Jerusalem's Dome of the Rock, with the eight entrances to its temple grounds adorned with pointed arches, Islamic architecture chose the ogival arch over the semicircular arches that dominated the Christian buildings of the late Roman Empire. Despite the religious differences, the pointed arch could rise high with a base of any span and thus was the perfect tool for the Catholic European masons when creating spectacularly high ceilings for their medieval cathedrals.

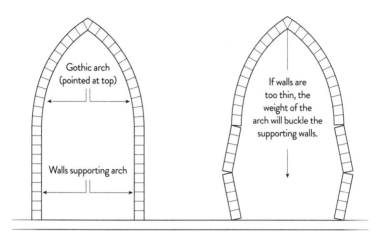

The most important aspect of designing a cathedral is sizing the walls that support the roof arches. If the walls are too thin, the weight of the arch will cause the walls to buckle, and the structure will collapse.

To safely and economically construct such breathtaking beauty, the cathedral's mason had to correctly size the thickness of the wall supporting the arch.[6] If this wall were too thin, the weight of the arch would buckle the wall. If the walls were too thick, stone would be wasted and the desired open space inside the cathedral diminished. To size the wall, the head mason would need to use a rule inherited from late antiquity, a rule that created the Pantheon and Hagia Sophia: a stable arch results when the wall's thickness is a little more than a fifth of the arch's span. The head mason, though, had likely never learned to read, let alone calculate a dimensional ratio. Instead, he carried out this rule without a yardstick marked with numerals, without Euclidean geometry—with only the most basic mathematics.

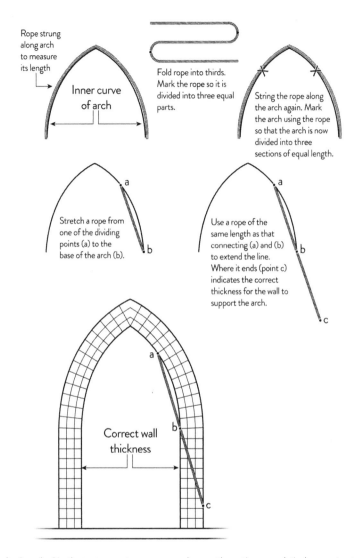

Rope strung along arch to measure its length

Inner curve of arch

Fold rope into thirds. Mark the rope so it is divided into three equal parts.

String the rope along the arch again. Mark the arch using the rope so that the arch is now divided into three sections of equal length.

a

Stretch a rope from one of the dividing points (a) to the base of the arch (b).

b

a

Use a rope of the same length as that connecting (a) and (b) to extend the line. Where it ends (point c) indicates the correct thickness for the wall to support the arch.

b

c

a

b

Correct wall thickness

c

As described in the text, a master mason used no mathematics or analytical geometry to design a cathedral. Instead, he implemented a proportional rule using a length of rope.

He ran a rope along the arch template, as if draping the rope over the arch itself. He then cut the rope to equal the full length of the arch as it curved from the first wall, up to the arch's peak, and down to the other wall. Then, laying the freshly cut rope straight, he folded it into thirds and marked each fold with colored chalk. With the rope now marked into three sections of equal length, he returned it to its original place draped along the arch template. Using the chalk marks on the rope, he could mark two key spots on the arch itself, each falling a short way down from either side of the arch's peak. By pinning the rope in each of those chalk-marked spots, he could then pull the rope's bottom thirds taut to create straight lines from the pinned spots to the points where each side of the arch met its supporting wall. The length of that straightened portion of rope and its particular angle became key, as the mason would then cut another portion of rope identical to it in length and lay it, end to end, to extend the path of the taut portion of rope in a straight line. That extension would become the hypotenuse of a right triangle in the mason's head—though it's unlikely he had heard of either in his life—the shortest leg of which would become his final measurement: the width of the arch's supporting walls, which ensures centuries of stability, all without even the simplest mathematical calculation.

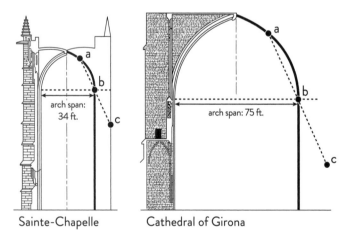

Sainte-Chapelle   Cathedral of Girona

(A) The Sainte-Chapelle, built in the mid-thirteenth century; (B) The Cathedral of Girona, built in the early fifteenth century. Both used the same rule of thumb to size the structure's supporting walls.

This proportional rule derived from a thousand years of application and refinement. As more structures stood with dimensions defined by the proportional rule, that rule would continue to be passed on orally and used repeatedly. We can see it applied in the mid-thirteenth-century Sainte-Chapelle in Paris and in the main façade of the Cathedral of Girona built some 150 years later and six hundred miles south. This rule was one of many that formed a complex body of knowledge known only to head masons—rules that drew on the intuition a mason developed over a lifetime of building. "Use your own good thinking," a head mason advised his son in training.[7]

A typical judgment was to assess the quality of the stone used for the blocks: Did it have seams of weaker stone? Did it crumble or crack when tapped by a hammer? Did a block left exposed to water for weeks deteriorate? If using excellent quality stone, they subtracted three inches from the wall thickness as calculated above, and if weak stone, they added three inches. A mason might also make modifications during construction. If he observed the stones move or shake, then he would replace the stones with different shapes. Or the head mason would inspect the dried mortar and look for cracks that indicated stresses, then reinforce the structure.

This *is* the engineering method: a process of methodical and actionable problem-solving, the force that has created the human world as we know it. And by observing how medieval masons in thirteenth-century France harnessed this force, we can create a definition of the method at its most elemental:

Using rules of thumb to solve problems with incomplete information.

This simple statement seems to promise nothing exciting, yet it enabled a medieval engineer, the head mason, to build intricate, stunning cathedrals without the technical knowledge we prize today—no understanding of the strength of a material, the precise stress and strains inside a block of stone, or even the mathematical ability to calculate a simple ratio. And even more astonishing, the mason had no idea that he was missing this knowledge. At no point was a mason saying to himself, "If only I could calculate the breaking strength of this rock." To imagine he did so is anachronistic, reading today's way of engineering onto the past.

The masons' ability to succeed despite the shortcomings we perceive highlights the importance of rules of thumb, that the engineering method is composed solely of experience-derived, provisional guidelines, none of which guarantee a correct answer yet when woven together create works of stunning utility, reliability, and beauty. It seems almost unbelievable that the application of the engineering method revolutionized the world, yet from the flint tools of our cave-dwelling past to today's digital marvels, engineers have used these rules of thumb to accomplish the greatest acts of human creativity, to push the limits of human ingenuity.

A rule of thumb, more formally called a *heuristic*, is an imprecise method used as a shortcut to find the solution to a problem. It is an idea so old and pervasive that practically every language seems to have its own corresponding term, uncannily following a theme of body parts: in French *le pif* (the nose), in German *Faustregel* (the fist), in Japanese "measuring with the eye," and in Russian "by the fingers."[8] All these terms express an imprecise method of guidance by common knowledge, a protocol of estimation. By definition, a rule of thumb features "relatively unstructured methods" to achieve results.[9] In practice, it is *anything* that can plausibly aid the solution of a problem but is not justified from a scientific or philosophical perspective, either because it doesn't need to be or simply because it can't be justified through anything other than results. But these dictionary definitions lack punch and clarity. Although it's insightful to know the term derives from the Greek verb *heuriskein*, which means "to find," it's better to list the four key characteristics of a rule of thumb, which can be illustrated with a simple rule of thumb used to improve a player's chess game: "Control the center of the board."

First, a rule of thumb reduces the time and effort needed to search for a solution to a problem. A player could plan for as many specific game scenarios as possible, but by generally positioning their pieces to cover spaces in the center of the board, most of those scenarios will be covered without fretting about the details.

Second, it can secure a probability of success within given parameters, but it does not guarantee success. A player who controls the center of the board will not necessarily win every game, but a chess hobbyist who makes a point of doing so will be more likely to win against opponents who ignore this rule.

Third, it can remain valid while simultaneously contradicting other rules of thumb that help solve the same problem. A player may also win by remembering to "establish outposts for your knights" or "keep your bishops on diagonals," even while doing so could give up control of the board's center.

Fourth, it rejects absolute standards. Rules of thumb are designed to be applied and judged according to a problem's context but become less useful, perhaps even meaningless, when considered abstractly or objectively. A chess theorist won't be able to find solid grounds on which to say that "control the center of the board" is any better a rule of thumb than "save your king's moves for the late game," and a player might find that every rule they followed before suddenly becomes useless when they try a game of speed chess.

Apply these characteristics to the cathedral's head mason. First, the proportional rule could size a stable wall width in a matter of minutes without spending the time needed to learn the mathematical knowledge to which he lacked access. Second, although any grand stone structure ran the risk of collapse, he could be certain enough that masons throughout history had supported a cathedral with arches designed using the proportional rule, and his was likely to survive as well. Third, sizing a supporting wall using this rule alone might create too weak a wall, so as the head mason determined the stone's quality, he invoked other rules of thumb that altered the calculation. Finally, these Gothic design rules that are so robust when building cathedrals fail absolutely and catastrophically when applied beyond the construction of medieval stone architecture. Modern methods reveal the heaviest load that a particular structure can hold to an engineer, who then designs the structure with a safety margin that holds up to five times the expected load. If instead an engineer decided to bring back the mason's proportional rule when designing a modern skyscraper, it would collapse into rubble under its own weight likely before even being completed. Yet many of the mason's cathedrals that follow this same proportional rule survive today as houses of worship and tourist attractions.

It's tempting to think of the methods of antiquated builders as placeholders until the "real" answers arrived in our scientific age to build our skyscrapers and spacecraft. This might lead us to view the mason's design methods as "protoengineering," a primitive method that evolved into the sophisticated ones used by modern engineers. But nothing of the kind happened. These proportional rules worked in Gothic buildings because the masons never exposed their available and abundant material, stone, to strain anywhere near its failure point, although they never knew this! A stone column will not crush the stones at its base until it reaches 6,500 feet—far higher than the tallest cathedral of the Middle Ages, the 404-foot-tall spire of Salisbury Cathedral.[10] In their time and place, the mason's rules of thumb were indispensable and insurmountable, but once no longer useful, the Gothic design rules, instead of evolving, just disappeared, vanishing both from use and from memory thanks to the oral traditions of medieval apprenticeship. Only in the last fifty years have architectural historians cobbled together these rules using fifteenth-century pamphlets and reverse engineering based on measurements of the cathedrals themselves.

That rules of thumb can be tossed out as easily as they coexist draws a contrast between the scientific and the engineering methods. The value of a rule of thumb isn't established by conflict as in scientific theory. Think of Einstein's theories replacing those of Newton. Newton visualized space and matter as absolute and fixed—a rigid object moving in space remains the same length. Einstein showed, however, that as an object moves, it contracts in length. Given the opportunity, we could see this phenomenon when we are moving near the speed of light, but while we are cruising down a highway, the complex equations of Einstein become identical to Newton's simpler equations of motion. With this discovery, Newton's theory was proven wrong and, although revered in history, abandoned by theoretical physicists. The medieval mason's proportional rule, meanwhile, was never proven wrong. The cathedral arches are still around today as proof in its favor. It was the material world—the development of iron and steel—that left the rule behind.

Put another way, the two methods have different goals: the scientific method wants to reveal truths about the universe, while the engineering method seeks solutions to real-world problems. The scientific method has a prescribed process that we all learn in school—state a question, observe, state a hypothesis, test, analyze, and interpret—but it doesn't know what will be discovered, what truth revealed. In contrast, the engineering method aims for a specific goal—an airplane, a computer, a cathedral—but it has no prescribed process. The engineering method cannot be reduced to a set of fixed steps that must be followed, because its power lies exactly in the fact that there is no "must." The specialized skill of an engineer is to find the correct strategy to reach a goal, to select among, combine, and create the many rules of thumb that will lead to a solution. Most often a rule of thumb results in a numerical estimate of some quantity, but it also guides an approach to problem-solving. A common one that illustrates the flavor of this class of rule is "break complex problems into smaller, more manageable pieces." Thus, the engineering method is best described as an attitude or approach or even a philosophy of creating a solution to a problem.

Consider a tiny object essential to the operation of airplanes, automobiles, tractors, or air conditioners—anything with motion. It costs fractions of a penny and has headlined in history only once: a thin, ring-shaped tube of rubber, the tiny yet mighty O-ring. It entered the vocabulary of every American after the *Challenger* space shuttle disaster of 1986, when physicist Richard Feynman dipped an O-ring into a glass of ice water to reveal NASA's failure to consider the importance of temperature. At low temperatures, the O-ring stiffened and failed to seal, which allowed exhaust to escape and rupture a fuel tank. The O-ring seems so simple and obvious that it couldn't have been invented. The intentionality of its design isn't evident, but there is a subtle trick to making an O-ring work, a trick discovered by Niels Christensen after forty years of thinking.

Christensen's technical training began in 1879 when he was fourteen and he left public school to apprentice at a machine shop in Vejle, Denmark, while at night he prepared to apply to the Polytechnic Institute in Copenhagen.[11] His industry and dedication worked: at age eighteen, he was accepted by the institute and placed in a role with the largest engineering project in Denmark, the Hanstholm Lighthouse. This lighthouse on the northern tip of the peninsula protected the entrance to the North Sea with a two-million candlepower light—so bright that someone twenty-four miles away could read by it, so powerful that every night, it stunned and killed scores of birds. Each morning, workers had to haul basketfuls of starlings, snipes, and larks from the lighthouse grounds.

When Christensen graduated with the highest honors, he enlisted in the Royal Danish Navy as an engineer, where he asked for an assignment abroad. The navy, recognizing his superb mechanical ability, granted him permanent leave and a stipend to work in the industrial powerhouse of nineteenth-century England. But from his new post, he could watch the rise of the United States as a rival power. Attracted to what he called its "mechanical progressiveness," he resolved to travel across the Atlantic, landing in New York in 1891.[12] He soon settled in Chicago, where he designed steam engines for Fraser & Chalmers as they outfitted the streetcars of the nation's rapidly developing transit system.

To learn more about the competitive market, Christensen took a short trip to inspect the new electrical railroad known as the Cicero & Proviso system in Oak Park, Illinois, where he witnessed a tragedy that would change his life. Shortly before 1:00 p.m., an electric streetcar barreled down the track toward a head-on collision with a Northern Pacific passenger train. The signalman furiously flashed "stop," but when the streetcar driver slammed on the brakes, they refused to respond. Streetcar passengers scrambled to the doors and jumped to escape the inevitable collision. The driver of the passenger train slowed his vehicle as much as time and inertia would allow, but still the streetcar was crushed against the Northern Pacific engine. The impact fractured the skull of the streetcar driver, killing him, and threw from the streetcar a passenger who later died in the hospital.

Inspection revealed that the electrically powered brakes failed when the car itself lost power—the brake shoes were helpless to press against the wheels. The problem was seemingly intractable. Stopping the movement of a massive powered mechanism would surely require an equally powered mechanism. Unless, Christensen realized, energy was stored and released to brake the car by something separate from the car's electrical power.

Obsessed by this problem, he devised an ingenious pneumatic system with a complicated triple-action valve that forced air into a cylinder and sealed it before the car moved. With the air stored and released mechanically, it didn't matter whether the electricity shut off while the car was traveling. The required pressure would always be available. For forty years, his braking systems were used by nearly every electric streetcar in service, but the complex problem and solution of sealing the cylinder never left his mind. In 1933, Christensen, then sixty-eight, still found himself working on sealing the fluid in brakes, but he thought his triple-action valve solution frustratingly inelegant for providing a pneumatic seal. If a reliable seal could be formed by a single, inexpensive part, creating lifesaving brake mechanisms could become trivial.

He and others had tried rings made of newly developed synthetic rubber to seal pneumatics, but they wore out quickly. This time, though, he had the thought to cut a groove into his pneumatic system's piston, slip the rubber ring over it, and pressurize the container. He found, as had others before him, that it failed. But choosing to follow his intuition, if only out of curiosity, he cut grooves with slightly different dimensions, returned the ring to its place on the piston, and slid the piston back and forth to examine the rubber ring and observe how it wore down. Each new groove slowly revealed the trick to the O-ring, just as trivial as the object itself: cut the groove to one and a half times the O-ring cross-sectional radius—a new rule of thumb! The result was remarkable. "This packing ring," he wrote in his notebook, "has been tested" nearly three million times and "has never leaked and is still tight."[13] The O-ring was so simple that no one thought it would work until two World War II Army Air Corps engineers used it to fix some leaking brakes on the landing gear of a Northrop bomber. Soon this simple but ingenious O-ring seal appeared everywhere, including fountain pens, soap dispensers, plumbing systems, hydraulic presses, washing machines, automobiles, and, most notably, spacecraft.

To Christensen's mind, the O-ring sealed because it was, as he said in his patent, "continuously kneaded or worked to enhance its life."[14] According to his notes, his subtle adjustments had unlocked the potential of rubber to behave like a strand of muscular fiber, growing in strength as it is exercised. Christensen was, of course, wrong. He never understood why the O-ring worked, and neither did anyone else until photographic techniques in the 1940s revealed the groove and ring's operation. Because Christensen cut the groove wider than the ring size, the ring rolled slightly when pushed by the moving piston. The rolling deposited a microscopically thin lubricating layer of hydraulic fluid between the rubber and the cylinder wall that hugely increased the ring's operational life span.

Christensen's invention leads us to a broader, richer definition of the engineering method. He wanted to *change* something in the world: to prevent brake failures. His options to effect that change increased over time; the types of materials available changed, but more abstractly, his *resources* increased. His first solution, that complicated triple valve, drew on late-nineteenth-century state-of-the-art machining. His second, the O-ring, used newly developed synthetic rubber. And even though he couldn't provide an accurate explanation of why his solution worked and the ability to do so wouldn't exist for another decade, his intuition and diligent trial and error brought him to a solution despite his lack of *understanding*.

Given this more nuanced perspective, we can tweak our definition of the engineering method a bit to realize its more comprehensive form:

Solving problems using rules of thumb that cause the best change in a poorly understood situation using available resources.

This enduring philosophy of problem-solving provides an opportunity to uncover one of the most consequential forces behind the creation of the human world. These terms—*rules of thumb, resources, optimization, uncertainty*—like their predecessors in our original definition, do not seem to promise a world full of steel towers, lithium-powered cell phones, ocean-crossing airplanes, and spacecraft journeying outside our solar system. But they are all enriched by their synthesis into the most powerful problem-solving method we humans have brought to bear on our planet. Every human who has designed a solution to a complex problem has tested the limits of resources and strived to conserve them, defined the optimal state of a desired outcome, prepared for and carved paths around the inevitable uncertainties of the world we live in, and interwoven our shared inheritance of tried-and-tested rules of thumb that have been created, passed down, forgotten, and rediscovered for millennia of human history.

Calling the engineering method a "philosophy of problem-solving" is not a pithy or idle description. The method is itself an adjective describing an underlying mindset that manifests constantly in our everyday lives but is rarely articulated, most often invisible. With its articulation comes a deeper understanding of human nature and thought, as if the method is human nature itself forged into a tool to shape the world, a tool never noticed or appreciated on its own but always in use, like the sawdust-covered table in a workshop.

To see how we use this method to create the world, the words and phrases in its statement will need to be further defined, nuanced, and, especially, illustrated; as Bertrand Russell noted, "Nothing of any value can be said on method except through examples."[15] By reading this book, you will hopefully be able to come to a deeper understanding of this tool and the infinite forms it can and is meant to take, throughout history and in the lives we live today.

2
BEST

The Vitruvian Man, Kodachrome, and
the Constant Search for Better

Leonardo da Vinci's *Vitruvian Man,* the famous drawing of a human
figure inscribed in a circle, presents the Renaissance ideal of bodily perfection, spelled out with mathematical precision by da Vinci in a notebook: "From the roots of the hair to the bottom of the chin is a tenth
of a man's height; from the
bottom of the chin to the top
of his head is one-eighth of
his height; from the top of his
breast to the top of his head
will be one-sixth of a man."[1]

Although the beauty of
da Vinci's perfect Vitruvian
pleases our eye, its notion of
"best" is nearly the opposite of
an engineer's. Of more use to

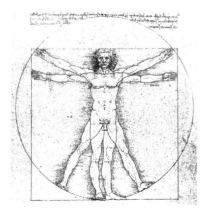

Leonardo da Vinci's *Vitruvian Man.*

engineers are the drawings of ordinary people, with their far-from-ideal proportions, developed in the late 1930s by industrial designer Henry Dreyfuss.

Dreyfuss wanted to make things for real people, and to do so, he would need to know what real people looked like. But lacking the funding or know-how to run his own anatomical survey of Americans, his firm culled anatomical data from the U.S. military for a sample of men's measurements and from high-end clothiers for women, then drew figures to represent the fiftieth percentile of all men and women. These data became a series of pricelessly useful drawings for any industrial designer of his time and afterward. Dreyfuss's *The Measure of Man* published his collection of averages with "Joe and Josephine" displacing da Vinci's single ideal. The pair illustrated the average American male and female, as Dreyfuss found them in the statistics, side by side with Joe at five feet, eleven inches and 162 pounds and Josephine standing five five and 135 pounds. Da Vinci might have preferred that the height of a man's head be 10 percent of his body, but Joe's was closer to 13.

Dreyfuss used his booklet's charts of statistical heights, lengths, widths, and weights of every body part he could find to create, as he liked to say, anything that was "going to be ridden, sat upon, looked at, talked into, activated, operated, or in some way used by people."[2] John Deere tractors, a jetliner cabin, and the inside of *Time* magazine all came from these measurements, but he also used Joe and Josephine to be sure the mundane fit into our human world: shaving-brush handles, perfume bottles, belt buckles, and neckties. He demonstrated this sensitivity to human needs most clearly in two of his most ubiquitous designs: the iconic model 302 telephone where, to size the handset, he used the measurements of two thousand faces to estimate the average spacing between mouth and ear, and the round Honeywell thermostat, which perfectly fit the typical human hand.[3]

When creating the "best" phone or thermostat, there is no such thing as designing to an absolute standard, no ideal of a Vitruvian Man. Instead, there is only a relative standard: a woman's typical hand to size a brush handle, an average size to serve the most users for a telephone handset. Dreyfuss's work leads us to what *best* means in engineering: the best-engineered design emerges from juggling hundreds of restrictions and parameters without which there is no such thing as the best version of anything. An engineering solution can only be judged based on how it handles the constraints unique to its situation—a balance of cultural forces, societal values, availability of material resources, and even urgency. The best solution is best "all things considered," because an engineer creates in a culture, not in a vacuum. A telephone is used by a Joe or a Josephine and not by the Vitruvian Man.

This engineering meaning of best becomes most apparent in a cross-cultural comparison of solutions to the same problem. When needs held in common by a society are transplanted somewhere else, with different resources, innovations, and scale, it is the society that allows us to judge the solution, not the problem itself. This point is illustrated by two different technologies developed by the Egyptians and the Arawak—the former to create ceremonial wine, the latter to remove poison from their most important food source.

Section of a hieroglyph at the Fifth Dynasty tomb of Iymery at Giza that shows the extraction of wine from grapes or oil from olives or other plant matter. Although the workers on the left are obscured, the workers in the middle and on the right can be clearly seen.

In Egypt, the regular flood season of the Nile deposited rich silt along its banks, which became fertile farmlands that sustained a population of about a million, about one-tenth of the entire world's population at Ancient Egypt's peak. In these fields, agrarian workers, under a highly organized annual regimen, grew grain for bread and beer, flax for linen, papyrus for paper, and grapes for wine. The last was produced for funerary rites of royalty and nobility who retained the honor of passing into an exalted afterlife. To extract the juice from grapes, the Egyptians used a simple but effective process that was essentially wringing out a cloth. They filled a cloth sack—probably linen—with grape pulp, sewed the sack shut to encase the pulp in a long tube with poles threaded through both ends, then hoisted it into the air like a snug hammock. A five-man team—two workers on each end and one in the middle—manipulated the sack to squeeze juice from the pulp. The man in the middle suspended himself in the air horizontally between the rods to push them apart with both hands and feet from the top as the workers at the ends twisted the sack. The torsion squeezed out the grape juice, which flowed through the cloth into a large pot beneath, in which it would ferment into the wine appropriate for sending the highborn to Anubis.

This technology served the Egyptians and the other civilizations of the Mediterranean well for centuries. That longevity and popularity might argue for the stretched torsion cloth method as the best technology in an absolute sense, yet the Arawak of the Amazon River developed a different technology for separating liquids and solids that is a threatening contender for best.

In Brazil, where the Amazon and the Rio Negro meet, a "blackwater" river forms. Its slow-moving channels of coffee-colored water nourish vibrant green banks that support the most biodiverse region in the world. This tropical rain forest has for eons been a magnet for human settlement, often pictured in error, wrote an anthropologist, as filled with "stone age savages frozen at the dawn of time" who lived hand to mouth in small villages in the pristine forest.[4] This characterization is driven by what the region lacked: stone architecture, writing, grain surplus, and domesticated ungulates—the hallmarks of civilization as measured by the standard of Ancient Egypt. Recent research, though, reveals that a large, complex society existed two thousand years ago across Amazonia, with dense, permanent settlements along the major waterways and a population of five to six million by some estimates.[5] The major group in this population was the Arawak, the largest language group in the pre-Columbian Americas, who transformed the inhabitants of Amazonia from small-scale foragers to farmers who cultivated soils to increase yields. These farmers created a dark, rich earth called *terra preta*, Portugese for *black earth*, by mixing in plant refuse, fish bones, and the charcoal debris from the slow-burning fires used to clear the forest. In this soil, rich in potassium and phosphorus, they grew their staple crop, manioc.

Manioc, a tuberous root similar to a sweet potato, is a source of carbohydrates and is still the fourth most important dietary source of energy in the tropics. Manioc was key to the Arawak's survival, which compelled them to develop sophisticated methods to process it. To stop the decay of tubers after only a few days, they turned them into shelf-stable flour used to cook flatbreads and thick gruel, the latter boiled with a fruit called *iriwa* to give it a prune-like taste. Manioc sustained the Amazonian people, feeding the millions spread along the banks of the second longest river in the world, all despite the single caveat to the plant's millennia of utility: manioc contains cyanide.

Once manioc is harvested, the tissues break down and release the cyanide trapped within its tissues as hydrocyanic acid, also called prussic acid. Ingesting the acid induces vomiting, dizziness, paralysis, and death. The Arawak farmers knew of less poisonous varieties—described as "sweet" in contrast to the poisonous "bitter" variety used—but the poisonous manioc had a greater yield, resisted insects, tolerated acidic soils, and had a higher carbohydrate concentration.[6] These are all fantastic properties for any staple crop but hardly useful if the food kills anyone who eats it, so the Arawak had to detoxify it.

When the poisonous hydrocyanic acid boils a few degrees above water, the acid evaporates. Heating, though, destroys nutrients, mars the texture of the manioc, and required villagers to spend extra time gathering firewood. But when the manioc is squeezed or crushed, the toxic acid flows from the manioc as a white, milky liquid. So to detoxify manioc, the Arawak faced the same problem as the ancient Egyptians: how to separate solids and liquids. To separate the milky liquid from the solid on an almost industrial scale, they created the *tipití*.

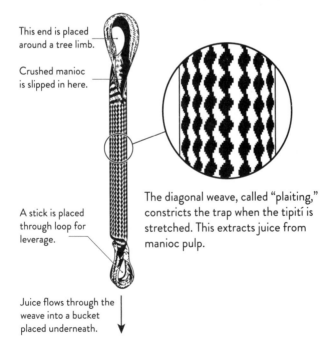

This end is placed around a tree limb.

Crushed manioc is slipped in here.

A stick is placed through loop for leverage.

The diagonal weave, called "plaiting," constricts the trap when the tipití is stretched. This extracts juice from manioc pulp.

Juice flows through the weave into a bucket placed underneath.

To detoxify manioc, Arawak farmers used a device called a *tipití*.

A tipití most closely resembles a Chinese finger trap—the novelty toy that tightens around a victim's fingers as they try to remove them thanks to a diagonal weave, called "plaiting," that constricts the trap when stretched—only the tipití is a woven tube about seven feet tall with a loop at each end.[7] To use a tipití, a farmer compressed the ends to open it wide, then filled it with manioc pulp created by a grater studded with some two thousand fine, sharp bits of palm wood. The farmer looped the top of the tipití around a horizontal pole secured between two trees, then slipped a long tree trunk through the bottom loop. Then, with one stomp on the trunk to pull the weave tight and constrict the tube, just like the finger trap tightening, juice flowed from the pulp through the plaited mesh. The few seconds it took to stomp on the trunk contrasts with the hours that would be spent collecting firewood if the Arawak had tried to boil off the hydrocyanic acid. The tipití removed the bulk of the liquid, and then gentle heating removed any poison that remained.

In comparison to the tipití, the Egyptian twisted sack seems crude, a flimsy bag whose use is labor intensive compared with an elegantly pleated, one-person-operation tipití. Anyone is welcome to appreciate the tipití for its intricate design, but the answer to the question of which one was better or best is "both."

Three major reasons illustrate why each solution was the best for its time and place. First, the labor supply differed between the two civilizations. The Egyptians were squeezed into a small region around the Nile, and their large agricultural operations used only a few workers because it was concentrated, freeing a workforce in the tens of thousands for other projects. A five-worker team to extract wine, a rare luxury in the Old Kingdom, was no great expense to the nobility who wanted it. In contrast, the Arawak were labor poor; families tended small manioc plots dispersed through the rain forest.

This motivated the Arawak to find a labor-efficient method like the tipití to process their dietary staple.

Second, each design drew on existing technologies widespread in the civilization. The Egyptians excelled at weaving cloth. Laundry scenes carved on tombs and laundry lists inscribed on pottery shards reveal cloth was used for dresses, towels, bedding, cushions, bags for carrying spices, and lamp wicks. The carvings illustrate the spinning of flax into the linen central in Egyptian elite funerary practice. After water was leached from the corpse's organs with natron, a mineral salt, and allowed to dry, the facial orifices were plugged with cloth and any cavities in the body filled with the linen. Then, the body was bandaged, first fingers and toes, then arms and legs, with any missing limbs replaced by a roll of cloth, then a final wrapping of the whole body. In Egypt, there was a ready supply of cloth with the right tensile strength and porosity for use in separating grape pulp and juice.

Although the Egyptians mastered the weaving of cloth, their methods seem simple when compared to the sophisticated weaving at the heart of Arawak life. The anthropologist Claude Lévi-Strauss observed that the modern descendants of Arawak lived in houses that "were not so much knotted together, as plaited, woven, embroidered."[8] Even if the Egyptians had conceived of something like a tipití, their weaving techniques using coiled fibers were not up to the task. The ancient Egyptian method requires no special workshop or skills, only an awl or a needle and perhaps a knife, but it is labor intensive, and a coil tube could never work as a tipití. It couldn't handle the strong pull that plaiting could sustain thanks to the combined strength of strands interlaced with crossing strands.

And finally, the third major reason that each solution was best: the abundance of particular materials, a design constraint we'll examine in detail elsewhere. The Arawak wove with a tough, strong fiber stripped from the herbaceous plant *Ischnosiphon arouma*. In contrast, the Egyptians' main material for weaving was the more delicate palm leaf, so even if they had the technique of plaiting, their materials were substandard compared with those of the Arawak.

When we say that best, for an engineering solution, can only be judged based on its response to the constraints from material resources, societal needs, and existing technologies, then we are also saying that best inextricably comes from culture. On the upside, this impulse to create, to solve no matter what, empowers engineers to use their imprecise, sometimes uncertain rules of thumb to aid society—to invent the lifesaving pharmaceuticals, safe cars, and sturdy buildings we cannot wait for while philosophers, scientists, and mathematicians debate truth, nature, and certainty. This best, though, is a double-edged sword. When best is relative to cultural goals and values, no technology is neutral.

A technology carries with it the choices and biases of the engineers who created the technology. In our modern postcolonial era, those have most often been dictated by white middle-class men. Crash test dummies were originally based on the average male, so the results of tests simulating frontal impacts—ramming a tree, T-boning other cars, or squarely hitting a wall—were ineffective for women.[9] Although today's automakers can use a "Hybrid III" family, which includes a woman, three children (ten-year-old, six-year-old, and three-year-old), and an oversized male cousin, many countries, including those using European Union regulations, require tests with only an average male dummy. Other, less lethal forms of engineered bias include the algorithms used to control temperature in large office buildings. The formula, developed in the 1960s, set the temperature based on the metabolic resting rate of the average man; these offices chilled women, whose metabolic resting rate is 35 percent lower. Even a fifty-fifty division of toilets in an office build-ing, something written into most building codes, results in inequal-ity, because women on average take 2.3 times as long to use a toilet as men. The built-in bias extends beyond gender: voice-recognition systems often only recognize white, male voices. And many designs embed ableism: most game controllers require two hands. The algo-rithm that drives search engines like Google struggles to serve people who are not white and male. One professor of information sciences searched "Black girls" and discovered, to her horror, that the top results led to porn sites.[10]

The implicit parameters these solutions work within allow engineers to judge their solutions as a kind of best as it has been defined. But in the pursuit of that best, engineers often uncon-sciously skip past the step of judging the parameters themselves. In part, this is because a historical lack of diversity, both in race and gender, in professional engineering limits engineers' perspective: those who haven't lived as a woman aren't likely to realize that a par-ticular temperature feels colder to a woman. But the sizable remain-der of the blame goes to an unconsciously narrow perspective of whom the solution will help or who will benefit from a product. If car makers thought that mostly men drive cars, then the best safety solutions were ones that prioritized protecting men. If search engine programmers believe that mostly white men are looking for infor-mation on the internet, then ignoring the interests of young Black women doesn't diminish their solution's effectiveness, according to their parameters. One of the best examples of this bias is in two different methods used to create color photographs. The first was developed in the late nineteenth century and presented to a London audience at the Royal Institution in 1896.

A few minutes before nine in the evening, in April 1896, an audience settled in their seats at the Royal Institution in London's dingy but historic lecture theater. That night, some nine hundred people crammed themselves inside, hurriedly checking coats in hope of finding a seat in a theater that safely held only half that many. The large audience was attracted by the two-word title of the lecture, "Color Photography"—the decades-long dream of photographers since the invention of the technology.[11] Most in the audience had not seen the photographs of this inventor but had read gushing press reports: "colorific splendor" said one newspaper;[12] "their only fault being that the colors are more brilliant than in nature," said a prominent scientist.[13] The details of the method used to create these vivid colors were to be revealed that evening by its inventor, Gabriel Lippmann, a member of the august French Academy of Sciences.

At 9:00 p.m. sharp, Lippmann, a quiet, balding man with a neatly trimmed small beard and mustache and pince-nez eyeglasses perched on his nose, was introduced. As he approached the lectern, an opaque projector flashed his colored images on a screen: a stained-glass window, a woman sitting on a chair on a lush green lawn backed by vivid trees, a sunlit house covered in climbing plants, a still life of flowers and fruit, and a vibrant parrot. The few in the audience who had seen Lippmann's world-famous photographs—either by visiting exhibitions in Paris or five years earlier at a London-based exhibit at the Royal Society—knew that the projected images only hinted at the vivid colors when the photographs were viewed in person. The soft-spoken Lippmann opened his lecture by describing these color photographs as "*ma petite découverte*"—"my little discovery"—then continued in English, apologizing for "any shortcomings as regards the English language."[14] He tantalized his audience with how simple it was to create these images. "The photographs you are seeing," he noted, "needed approximately one minute of exposure to sunlight." He added that most of the photographs "were developed on the mantelpiece of a hotel room, which proves that the method is easy to carry out." He developed the photographs in only a "quarter of an hour" with no special equipment beyond the standard kit used to develop black-and-white photographs.

By noting this, he reminded the photographers in the audience of the complexity of previous attempts at color photography. Earlier color photos required specialized cameras that photographed an object through red, green, and blue filters, then dyed the resulting three negatives with their corresponding colors and combined them to generate a colored image. He called this an "indirect method" and dismissed it because the colors from the dyes were "in some degree [an] arbitrary choice" that did not replicate nature. In contrast, his photographs created colors without dyes or any complicated chemistry. They only needed, he said, the "direct action of the luminous rays."

To illustrate this, he reminded his audience of color produced by nature. "We all know," he said, "that colorless soap-water gives brilliant soap-bubbles"—referring to the rainbow of colors on a bubble's surface, similar to the subtle colors of a pearl and the shimmering colors of a peacock feather. These colors are created from colorless substances: soap water, calcium carbonate, and dull brown feathers. The color in these items, he explained, comes from closely spaced colorless layers that, when viewed from the correct angle, diffract white light like a prism, as first demonstrated by Isaac Newton. Lippman told his audience, "when two parallel reflecting layers are separated by a very short interval, and illuminated by white light, they reflect only one of the color rays which are the constituents of white light." Newton, Lippmann explained, showed that a spacing between layers of two ten-thousandths of a millimeter produces violet colors, and a spacing of three ten-thousandths produces red. In all three examples—bubbles, pearls, and feathers—nature creates color using only closely spaced layers of colorless materials.

The colorless materials Lippmann needed to produce color himself already existed in silver bromide, the chemical that enabled grayscale photography. When exposed to light, an atom-sized region of a silver bromide grain transforms into metallic silver. When developed, the specks of silver conduct electrons from a chemical developer to the bulk of the silver bromide, which grows into solid grains of silver, forming the dark sections of a negative. Lippmann's genius was to organize the silver in film into the thin layers needed to refract white light into colors.

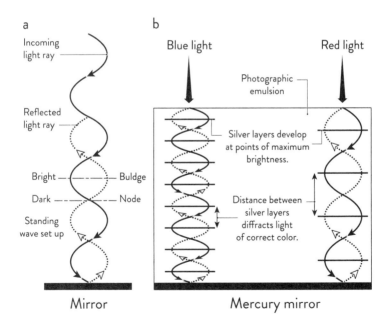

An incoming light ray and its reflection will form a standing wave. At some spots, the waves cancel each other, which results in a dark spot; in others, they reinforce each other and create a bright spot.

Lippmann coated a glass plate with a thick gelatin layer, called an emulsion (as one of his successors noted, the best choice was "a department store gelatin recommended as the best for puddings"[15]), mixed with grains of silver bromide. He then placed the plate in a camera and poured a thin layer of mercury behind it. When the plate was exposed to take a photo, the light passed through the emulsion to the mercury in the back, which acted like a mirror. As incoming light rays reflected off this mirror, the rays and their reflections combined to create waveforms with stable peaks and troughs, called a "standing wave." These standing waves were suspended, in a sense, in the photographic plate, where the peaks would activate the silver bromide as expected, while the troughs would leave the solution dark. In this way, the standing waves formed the microscopically spaced layers of silver that would refract colored light at the same wavelength as the light that shone on them when the photo was taken. The result was a plate of glass that used the physics of light to produce color from colorlessness, as with a soap bubble, pearl, or feather.

Lippmann's color photographs were a marvel of their time. They dazzled viewers with realistic human skin tones and subtle reflections from metallic objects that resulted from reproducing the entire color spectrum, which created a true white in the image that was impossible with the indirect three-color method. For this achievement, Lippmann won the 1908 Nobel Prize in Physics, still the only prize awarded for photography. As far as faithfully reproducing color in a photograph, Lippmann couldn't be beat by any of his contemporaries. But when judged by the parameters of his society's values and needs, Lippmann's method was a failure. Few manufacturers of photography apparatus commercialized his method, and it soon vanished from the marketplace.

Three fatal aspects doomed Lippmann's method to fall short of the criteria for the best solution to photography's problem. First, the photographs were difficult to view. To prepare the photograph for viewing, Lippmann coated the back of the glass plate in black paint and cemented a prism to the front. For the full colors to appear, a viewer had to be directly in front of the photograph; even five degrees off the correct angle caused a color shift—at some angles, a black-and-white negative image, at others, an image in unnatural shades of red and green. So if a group tried to view a photo, only one person at a time could see it at the correct angle. And any light illuminating the photo could only come from the front. Any light from behind would destroy the image. One onlooker who viewed backlit Lippmann photographs said, "it resembled a piece of London brown fog stuck on a piece of glass."[16] Second, photographers didn't like mercury's toxic fumes and the necessity of cleaning the mercury often so its mirror-like quality wouldn't disappear. And third, the method didn't fit with the times, because it took at least a minute, perhaps longer in low light, to capture an image. This, in the late nineteenth century, was far too long for a generation who had just moved from candles, horseback, and letters to electric light with the flick of a finger, rapid transport by car, and instantaneous communication by telephone. Today, Lippmann's photographs are museum pieces, their colors still brilliant and vivid because they used no dyes that could fade. But while Lippmann's color process slipped from memory, one of the great pioneers of photography—the man who transformed black-and-white photography from an elite hobby to an inexpensive activity for all consumers—searched for a better solution in the method dismissed by Lippmann, the indirect method that captured the red, green, and blue hues of a photographed object separately and combined them.

In the early twentieth century, George Eastman, the founder of Kodak, traveled the world in a quest for the best technology to create color photographs. Eastman visited professional photographers, scientific laboratories, and amateurs, looking at and assessing their solutions for color photography. "During my twelve weeks in Europe," he wrote in April 1910, "I spent a good deal of time on new developments in color...which I hope will develop into something commercial."[17] But none met the standard used by Eastman for all Kodak products: simplicity, a notion that turned Kodak into a tech giant, the Google or Apple of its day.

Kodak's first cameras, which used film instead of messy glass plates, stressed simplicity—"You Press the Button, We Do the Rest." Their revolutionary cameras and rolls of film spread around the world and penetrated markets no competing product had ever reached. In 1904, when the Dalai Lama came down from his Tibetan capital for the first time, he carried his Kodak camera.[18]

In 1917, a few years and most of a World War after Eastman's quest, two teenagers, Leopold Godowsky and Leopold Mannes, watched a blurry color presentation from a magic lantern, an early type of projector, in a New York theater.[19] The young men had formed a friendship through their shared interest in amateur photography and music, using Kodak's Brownie camera to snap pictures while also attempting to live up to the standards of their prominent musical families. The crude color film sparked their ambition to make something more impressive—an ambition that lies on the border between youthful exuberance and arrogance—but they wouldn't get a chance to fully realize their goals until they had graduated from college. Godowsky became a solo violinist for the Los Angeles and San Francisco symphony orchestras, Mannes had written a piano suite, and both had studied physics.

They each returned home to New York City to work as professional musicians but continued their experiments with color photography. Using a loan from their parents of $800, they converted their parents' kitchens and bathrooms into laboratories and darkrooms.

Early attempts included splitting the light entering the camera three ways through red, blue, and green filters to expose three different layers of film treated to react to the different wavelengths of light, which could then be combined to form a single, colored image. The three negatives, though, were difficult to register. The thickness of the negative "sandwich" degraded the image by filtering out so much light that the photo turned out dim and blurry because the thick layers scattered the incoming light. Their experiments failed, but their failures led them to the right solution: do everything on a single sheet of film.

To continue, they needed patronage that their parents were both unable and increasingly unwilling to give as the makeshift laboratories continued to take over their homes. To get the money they needed, Godowsky and Mannes went straight to the top, miraculously securing a meeting with Eastman himself. And although their pitch didn't secure any commitments from the owner of Kodak, it did eventually give them jobs with the company thanks to the impressed research director. He paid them $30,000 up front (about $500,000 today) for royalties on all patents and yearly salaries of $7,500 ($113,000 today). Their productivity—forty patents in their first year at Kodak—earned them the nicknames "God and Man," although some called them "those musicians" because of the whistling from their darkroom.[20] Godowsky and Mannes refused to wear watches with luminous radium dials, worried that they might fog the film, so instead they whistled Brahms's C-minor symphony to time the development of their test films.

Godowsky and Mannes's idea to solve their three-film problem was to instead coat a single sheet of film with three layers of emulsion, each layer being sensitive to either red, green, or blue light. This solved the first problem but created a second, intractable one: how to color each of these layers correctly when working with messy emulsion. When the film was developed, the colors bled from layer to layer, destroying the image. To reduce the bleeding of color between layers, the film would have to contain no dyes—exactly Lippmann's goal when he invented his refraction method of color photography. But Godowsky and Mannes would not do away with dyes entirely. Kodak chemists, at their direction, perfected developers that carried the dye into the film during the development and developed only one layer at a time—the red, blue, or green sensitive layers—and did not diffuse or move from layer to layer.

By the 1930s, Godowsky and Mannes, with technical support from Kodak, had perfected their film. It captured sharp, vibrant images. With retail prices of $3.50 for eighteen exposures including mail-in processing, it opened the world of color photography to amateurs, and within three years of going to market, Kodachrome outsold black-and-white film.

The new film fit within the parameters that Lippmann's method failed to consider. Lippmann's vibrantly and accurately colored refraction may have been more impressive to look at, better in terms of the absolute goal of transcribing the world's colors onto an image, but Kodachrome delivered on what turn-of-the-century Americans had come to value more: convenience. To do this, Kodachrome film transposed the complexity and simplicity of Lippmann's process: Lippmann's color photographs were complex to set up, with long exposures and toxic mercury, but they were simple to develop, a mere "fifteen minutes in a hotel room," as he noted. In contrast, taking a picture with Kodachrome film was a simple snap with an existing camera, followed by a Byzantine development process that could be taken off the photographer's hands.

For seventy-four years, Kodachrome vanquished all other color film to small markets until it was itself deposed by digital imaging. Kodachrome documented Queen Elizabeth's coronation in 1953 and, also that year, traveled to the top of Mount Everest, where Edmund Hillary snapped a photograph of Tenzing Norgay at the summit. It captured the piercing green eyes of adolescent Sharbat Gula, the "Afghan Girl" featured on the cover of *National Geographic*.[21] The photographer who captured this image described the color of Kodachrome as "a poetic look, not particularly garish or cartoonish, but wonderful...the gold standard of imagery."[22] The film was often celebrated for its "fidelity," as the *British Journal of Photography* wrote, and, as one enthusiast described it, "more like looking out a window at reality."[23] Fidelity to reality, though, was not a hallmark of Kodachrome.

By the 1950s, complaints about Kodachrome trickled into Kodak from their large corporate clients. A national candy company noticed that the film failed to capture the subtle brown hues that differentiated dark, bittersweet, and milk chocolate, and a furniture manufacturer reported that maple, oak, and other dark woods all looked the same. And occasionally, Kodak heard what many professional photographers had known since the late 1950s: that when taking class pictures of dark-skinned children, a photographer needed to reflect light into their faces from a low angle to preserve details, and even greater finesse with lighting was needed when photographing a dark-skinned person and a light-skinned person together, or the details in the dark-skinned person would vanish, leaving only the whites of their eyes and their teeth.

Kodachrome color film, like any film that uses three primary colors to create other colors, cannot replicate all colors because of the physiology of the human eye.[24] In our eyes, three specialized light-sensitive cells called cones detect color. The ρ cones detect red-orange-yellow light, the γ cones orange-yellow-green light, and the β cones green-blue-violet light. This suggests an image created from the three primary colors of red, green, and blue as captured and separated in Kodachrome would match the response of the human eye, but no color activates only the γ cones. Blue light stimulates the β cones, red light the ρ, but almost any shade of green activates both the ρ and γ cones. Although the dominant response to green light comes from the γ cones, the simultaneous stimulation of the ρ or the β cones creates paler greens, which in turn tinges white with magenta. To rebalance the colors for a purer white, tricolor films increase the intensity of the dyes that create red and blue to match the perceived increase in green, which comes at the expense of distorting yellow and brown hues.

The Kodak engineers defined their parameters for achieving the best color balance results—they called her Shirley. This photograph of a Caucasian woman wearing a colorful, high-contrast dress was updated at least every decade and included with every *Color Dataguide* published by Kodak. They advised photographers to create images "similar to the enclosed standard."[25] Those with automated printers used the negative of Shirley to set their equipment and achieve what Kodak called "good color printing."

But the colors of these Shirleys were not faithful to the model's actual skin tones; rather, they were set by a team of judges.[26] The photograph of Shirley was printed with differing tones, balance among the colors, and shades—from too red or yellow to too blue and from too green to too pink—in subtle steps. These altered Shirleys were inspected by judges, who voted on the "optimal" color balance. A Kodak engineer involved in this judging noted that the print closest to "exact reproduction," that is, matching the model's true skin tones, was described unanimously as too dark, while the print with the greatest number of votes was, when compared with the exact reproduction, "quite pale."[27] For years, Kodak, the sole developer of Kodachrome until 1954, optimized the processing to create pale Caucasian skin tones, leading to poorly reproduced darker skin tones.

The fashion industry embraced the vivid, vibrant tones of Kodachrome. Illustrative is the work of Louise Dahl-Wolfe, a pioneer in fashion photography. A photographer trained as a painter and adept at color theory, she latched on to Kodachrome soon after it debuted and created for *Harper's Bazaar* over six hundred color photographs and eighty-six covers between 1936 and 1958. "She wed," said a critic, "the American ideal of wholesomeness to a European standard of elegance."[28] These and other Kodachrome images in magazines like *Vogue, Glamour, Photoplay,* and *Marie Claire* helped create the beauty standard of the day.

This deficiency in color film reflects the fact that an engineer's notion of best is a choice embedded in the prevailing culture, that an engineered object, unlike a scientific theory, is designed to fit into the human world. Although this observation condemns the engineering method and its notion of best, paradoxically, it is only the engineering method that can remedy this unfairness, because an engineer's best is not an absolute standard—it changes with time. Acknowledging these built-in biases argues for a diverse workforce of engineers. The exclusion for centuries of more than half the population cuts across the spirit of the rules of thumb central to the engineering method: use everything that plausibly can lead to a solution. Every mind possible should be thrown at a problem. At its base, the reason for diversity in engineering is to increase the number of people whose unique knowledge might contribute to a solution, to even *notice* that a different solution is needed. The male engineers were oblivious to the poor match of standard bicycle designs to the typical shape of a woman's body. It took a female engineer to correct the built-in male bias.

Georgena Terry's path to a reengineered bike was neither straightforward nor motivated by a desire to correct an inequality. She simply wanted to use a blowtorch to create.[29] As is true of many engineers, her desire to modify the world around her started in child-hood. "I didn't hang out with dolls," she said. "Instead, I built sand-boxes and tinkered with tools." One project that took "endless hours" was lubricating the wheels of her wagon; she quested for the maximum speed down her parents' driveway. But as often happens with girls, pur-suing science and engineering is not reinforced throughout school, and with few role models, they drift away. Terry earned a degree in theater, but what fascinated her were the technical aspects—building sets and adjusting the lighting. Unfulfilled by this work, she enrolled at Wharton to earn an MBA, then worked as a stockbroker, but "being tied down to a desk" drove her "absolutely nuts." The desire to create in the physical world returned her to engineering, where Terry discovered her métier. How "comforting," she thought, "to work with engineering and science because they tell you the truth." She added, "You put real numbers in and get real numbers out."

As her final project in engineering school, she and a team built a hybrid between a bicycle and a car, a small vehicle that could be pedaled but had a small engine for power when needed. To assemble the metal tubing of the vehicle frame, one member of the team was adept at using an acetylene torch in a type of welding called brazing. This fascinated Terry. She peered over his shoulder as he brazed the tubing together and insisted that he teach her how to weld. After graduating, she worked a well-paying job at Xerox but only lasted two years before she quit to create bike frames for a living.

When she brought her homemade bikes to rallies, women would approach her and ask for a custom bike, because, they reported, standard bikes hurt their necks and shoulders. As a good engineer, she first quantified the problem: what are the anatomi-cal differences between cisgendered men and women? She turned to the work of Henry Dreyfuss to use the measurements of Joe and

Josephine. From Dreyfuss's careful work, Terry learned that a woman isn't just a smaller version of a man. A bike that fits a man's legs and upper body will not scale to fit a woman's dimensions, since a woman's upper body tends to be proportionally longer than a man's. This compounds with the fact that the center of a woman's muscle mass differs from a man's, stretching a woman's body as she rides a men's bike to cause neck and shoulder pain.

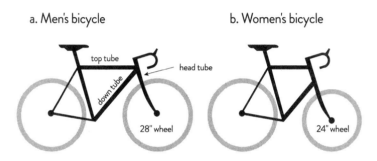

a. Men's bicycle   b. Women's bicycle

In a properly designed bike, the rider's weight should be split with 55 percent on the back axle and 45 percent on the front axle. (a.) A typical bike, one designed for men, shifts the distribution of weight for women toward the front axle because the distance between the seat and the handlebars is too long for them. (b.) To shorten the top tube so that the seat and handlebars are closer, a smaller wheel is used in front, which allows the down tube to slide down the head tube and opens space between the wheels for the pedals.

Terry redesigned the bike so that women users rode in a more upright position than on a traditional bike by shortening the distance from the seat to the handlebars and narrowing the handlebars to match a woman's shoulder width. She hauled her new prototype bikes to weekend biking rallies, allowing women to test-ride them. And from these first riders, Terry received what she called the finest compliment about her redesigned bikes: "Thank you," wrote one rider, "from the bottom of my butt." In her first year, 1985, she sold twenty of these women's bikes, the following year thirteen hundred, then five thousand. Her company, Terry Bicycles, became a multimillion-dollar enterprise serving the third of all bike owners and cyclists whose parameters had been previously ignored by the conventional "best" bike designs.

Although the examples I've discussed highlight the failure of engineers and the engineering profession to think broadly and deeply about all users, it is, paradoxically, only the engineering method that can resolve and prevent such blind spots in the future by modifying the toolkit used by engineers and, as you might guess, developing rules of thumb that guide engineers in designing for all people at all stages of life. This approach, variously called universal design, design for aging, and universal usability, features rules of thumb that help people as they lose range of motion, rules like "place cabinets three inches lower than the current standard," "replace knobs with D-shaped handles," "and use drawers instead of doors."[30] Yet universal design will always be at best an aspirational goal because it seeks an *ideal* solution. Ideals conflict with the notion of best as relative, a notion essential to the engineering method because it lets engineers create *a* solution *today*.

3
UNCERTAINTY

The Art of the Unknown and How
Engineers Use It to Create Solutions

In the late seventeenth century, French mathematician Jacques
Ozanam dreamed, in a literary compendium of twenty-three "useful
and entertaining" scientific problems, of a human-powered vehicle. "Its owner," Ozanam raved, "could freely roam along the roads
without having to care for an animal and might even enjoy healthy
exercise in the process."[1] For nearly a hundred years, a "horseless"
vehicle was dismissed as a pipe dream, until 1817, when the eccentric German baron Karl von Drais embarrassed his family, friends,
and colleagues by unveiling his *Laufmaschine*, or "running machine."
The device had two in-line iron wheels, a wooden carriage, a padded
seat, and a long, curved rod to pivot the front wheel. A rider pushed
off the ground with one foot, then the other, and at the moment of
Drais's first ride, the bicycle was born.

   In the last century and a half, engineers have refined the bicycle to be the most efficient vehicle in the world when measured in

energy expended per pound moved; it surpasses all moving creatures and machines, with a car using five times more energy by this measure.[2] With a billion bicycles on the planet, more than one hundred million manufactured each year, the bicycle is by all measures an engineering triumph, despite the fact that we have no idea how or why a bicycle works. "Everyone knows how to ride a bike," noted an engineering expert in sporting equipment at UC Davis, "but nobody knows *how* we ride bikes."[3]

When we learn to ride a bike as children, many of us have a naturally curious inclination to run a little experiment. Grasping the bike by its handle and seat, we give it a running push along the sidewalk to see how far it can go on its own before toppling over. The answer is an impressive distance, especially if it is rolling along a downward slope that allows gravity to accelerate the bike forward in place of the missing rider. By virtue of its design, a bike moving at a fast enough speed has an inherent ability to keep itself upright. And while you were learning to ride the bike, you personally got some help from this phenomenon as you started to pick up speed and confidence.

Yet this phenomenon is poorly understood, even though bicycles are a fixture in our world. Most engineers assume that, as with a gyroscope, the angular momentum of the front wheel helps the bike stay upright. But this seemingly obvious answer was ruled out in 2011 when a group of engineers designed a bike that specifically canceled out the effect with a suspended wheel rotating in the opposite direction to the front wheel, thus proving that the gyroscopic action of the front wheel was neither necessary nor sufficient for the stability of the bike.[4] If the answer wasn't a gyroscopic effect, then it could be in what the engineers call the "caster trail." Because the steering axis connecting the bike's handlebars to the front wheel's axle is designed to be angled backward, if the bike tilts to either side, the ground will push up against the front wheel behind the turning axis so that the handlebars naturally turn the bike in the direction it is tilting, keeping it from falling over. But the same team ruled out the caster trail explanation too. With a little creativity, it's possible to design a stable, self-righting bike that defies any simple explanation for how it does so.

A bike's stability seems to come from a complex mix of gyroscopic action, weight distribution on the frame, and the caster trail. But a mathematical definition of how these forces work together to produce the bike's stability has escaped us. "We have found," the team reported, "that almost any self-stable bicycle can be made unstable by misadjusting only the trail, or only the front-wheel gyro, or only the front-assembly center-of-mass position."[5] And conversely, unstable bikes can be made stable by adjusting any one of these parameters. We only need, then, to see a six-year-old zip down a neighborhood street on their bike to prove that engineering is not applied science, that a phenomenon need not be understood scientifically, mathematically, philosophically, or any other way before engineers use it to create something useful. The converse is also true: to design despite uncertainty signals that an engineer is at work. An essential task of an engineer is to maneuver their way around uncertainty, fence it off, then end-run past it to deliver results today, never mind how complex the phenomenon in question.

You yourself can create one of the most complex phenomena in nature by turning on your kitchen faucet. Adjust your faucet so water flows in a slow stream, place your fingers behind the pristine water column, and note how clearly they appear. Then open the tap and let the water gush from the faucet; the now opaque stream will obscure your fingers. The smooth, clear flow is called "laminar" flow, from the Latin *lamina* for thin plate or layer; the opaque, churning flow is called "turbulent," full of disturbance and commotion.

That transition from laminar to turbulent flow has puzzled scientists for more than a century. A physicist noted recently that "the central puzzle of why the transition takes place at all remains unresolved. It is an enigma... Despite wide-ranging research activity that has uncovered many important pieces of the jigsaw, the central puzzle remains unresolved."[6] It is so because scientists so far have no comprehensive theory of turbulence. There is nothing like Newton's laws of motion predicting the trajectory of a satellite or a bullet for fluid dynamics, because any comprehensive theory of turbulence must describe and predict a fluid's behavior from the atomic level to the macroscale. At the atomic scale, about 10^{-8} centimeters, sea water is composed of molecules that move according to quantum mechanical principles. At scales of a few centimeters to several meters, these molecules form waves. On a scale of thousands of kilometers, these waves form currents, and on a global scale, the tides that rise and fall with the moon. For most physical phenomena, one scale dominates, and scientists can develop a theory that predicts and explains how objects and forces behave. In the most complex phenomena, like the turbulent motion of water, scales across the spectrum, from atoms to oceans, are in play.

But engineers must know when the transition from smooth and placid to rough and violent flow occurs and how to control which type of flow manifests. Sometimes engineers need turbulent flow: the pharmaceuticals we swallow are mixed by turbulent flow, and turbulent mixing of fuel and air fires our car engines. In other situations, laminar flow is necessary: the steel for the buildings we work in is cast by the laminar flow of molten steel, and the silicon chips that power our computers and phones are prepared in a room swept clean by the laminar flow of air that removes chip-destroying contaminants. Nowhere, perhaps, is smooth flow more important than in the air over an aircraft wing. Without working around the transition between smooth and violent flow, "it is doubtful," said one engineer, "if practical aerodynamics could have proceeded at all."[7]

Although twenty-first-century science still cannot understand it, a nineteenth-century engineering professor used a time-honored strategy to enable engineers to work around the uncertainty of this extraordinarily complex phenomenon. In the last third of the nineteenth century, Osborne Reynolds taught at Owens College, a newly formed university in Manchester, England, where he had a reputation as a most peculiar thinker. His intellect lay in his deep, intuitive insight into physical phenomena, which he wrote about in an informal, concrete, almost naive language. "His brain," reported a former student, "seemed to work along lines different from those of the majority of us. He looked upon all things in an original manner," with an "almost primitive simplicity of mind."[8] He applied his powerful intellect to the natural world because the cash-poor university housed the engineering college in a stable next to a large hole fenced off by dilapidated railings behind their only building. Reynolds lectured on the ground floor of the stable and taught technical drawing to students in the hayloft. With no laboratory for research, Reynolds studied the tails of comets, the solar corona and aurora, the magnetism of the earth, and the electrical properties of clouds. He reported how hailstones, raindrops, and snowflakes form—even casting a piece of hail in plaster of paris so he could study its structure at his leisure. He investigated why trees burst when struck by lightning—the water inside expands when heated. He measured how sound bounces off the atmosphere and how fog attenuates sound. But of all the natural phenomena he encountered, none fascinated him more than the flow of water.

He puzzled about why sometimes it flowed calm and clear and other times violently—laminar and turbulent. He was partly interested in how understanding the phenomenon could be used for technical problem-solving—so obvious "as regards the practical aspects of the result," said Reynolds, "it is not necessary to say anything by way of introduction."[9] No doubt he was thinking of structures like the eighty-one-mile Dhuis aqueduct, which supplied Paris with millions of gallons of water from springs in the east of France for street cleaning, drinking, bathing, and, eventually, washing away Parisian waste in the most modern of sewer systems. French engineers designed, by arduous trial and error, the channels of this aqueduct so the water flowed smoothly; rough or turbulent flow would erode the channels and shorten the aqueduct's useful life. Reynolds himself remarked on the importance of laminar and turbulent flow in the efficiency with which ships move through water: as a ship's bow cuts forward through the water, the flow of water pushed to the sides is laminar, but as one walks along the deck looking down at the ship's side, at some point halfway to the stern, the flow churns into the turbulent eddies that trail in the ship's wake. "It is these eddies which account for the discrepancy between the actual and theoretical resistance of ships," Reynolds wrote.[10] Although the practical uses of understanding the transition sparked Reynolds's search for a method to calculate when and where laminar or turbulent flow would occur, it was the philosophical part—the puzzle—that captivated him. Entranced by the transition, Reynolds spent hours studying the "mysteries of fluid motion that had baffled" scientists.[11]

He wrote to a friend about conducting an experiment "on a windy day at Mr. Grundy's pond" near the college. He tossed a stone into the pond, tracking as best he could the motion of ripples, then poured a little oil on the pond and observed that the top of the pond no longer rippled with waves but "took on the appearance of plate glass," although beneath the surface, eddies of water still swirled.[12] Later, he watched rain fall on the pond and noted that the drops calmed the surface of the water. He continued the habit of hanging his head over the edge of ships to study the water. While idle aboard the ship, he studied the motion of waves and tried to follow a large wave, frustrated when it merged and vanished with other waves: "We find after following its course for a short distance that it has lost its extra size, while on looking back that this has been acquired by the succeeding wave."[13] Or he watched "the waves which spring from the bows of a rapid boat" and tried to trace the motion of a continuous section of water.[14] His inquisitive mind was frustrated that the motion was invisible. "Of the internal motions of water or air we can see nothing," he lamented.[15] "The exact manner in which water moves is difficult to perceive and still more difficult to define."[16] Then, as he continued his omnivorous observation of the physical world, he noticed smoke: "above a chimney, from the mouth of a gun, the puff of a steam engine, or the mouth of a smoker, or from Tait's Smoke Box."[17]

A Tait smoke ring box was a popular device that captured the imagination of the greatest thinkers of the nineteenth century.[18] The six-sided box had one side removed, which would leave it open except that it was covered with a tightly drawn towel; a hole an inch or two in diameter was drilled in the center of the opposite side. The owner of such a box filled it with ammonia and sulfuric acid to create smoke, then thumped the taut towel to blow large smoke rings through the box's hole. Lord Kelvin, after whom the Kelvin scale is named, studied the smoke rings as visual aids to develop his "vortex-atom" theory, which postulated that atoms were vortices in the ether.

Although Reynolds took no part in this theorizing, the smoke rings dazzled him. "The beautiful phenomenon of the smoke ring," he said, "the curls of smoke, as they ascend in an open space, present to the eye a hopeless entanglement."[19] He left for others the knotty problem of how to calculate the motion of these rings, but what dawned on him was that they revealed the "form of fluid motion."[20] Tait's smoke rings transformed the invisible air currents into the visible. This suggested to Reynolds a method to visualize the flow of water. At his next opportunity, he injected dye near the propeller of a ship and noticed, to his delight, "a beautiful vortex ring, exactly resembling the smoke ring."[21]

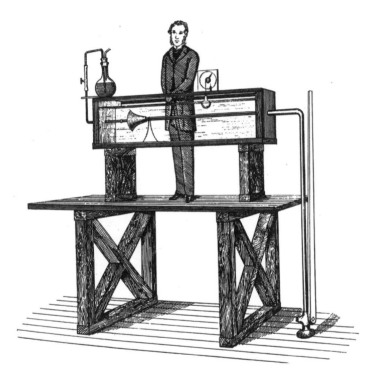

In the mid-nineteenth century, Osborne Reynolds built a special tank to study the transition from laminar to turbulent flow. When the valve near the floor was opened, gravity sucked water into the wooden funnel inside the tank and then into the glass pipe. As the water flowed, Reynolds injected a small quantity of dye from a nozzle to delineate the water's flow.

With this new technique for visualization, Reynolds designed an apparatus to learn the mysteries of fluid motion, to predict when and, he hoped, how the transition from smooth to violent flow occurred. He built a wooden box six feet long and a foot and a half in width and breadth, with the front and back of the box being clear glass plates.

He placed this box on a platform about seven feet above the ground and filled it with water. Inside this tank of water, he suspended a five-foot-long glass tube one inch in diameter fitted on one end with what he described as a "trumpet mouth of varnished wood"—we might call it a funnel.[22] He slid a sleeve of india rubber on the other end of the glass pipe to attach an iron pipe, which passed through the wooden end of the box, bent downward ninety degrees, and traveled seven feet to a drain in the floor with a valve controlled by a large lever.

With his device ready, he could begin his experiments to discover one of the key mysteries of fluid dynamics. He filled the tank with water, let it stand for several hours to ensure that the temperature throughout was as close to uniform as possible, and cracked open the valve slightly. Inside the wooden funnel, he had placed a tiny glass tube, so small that it would not disturb the flow, through which he would then release dye into the water. As the water flowed, a single, unperturbed line of dye floated through the center of the larger tube "and remained beautifully steady."[23]

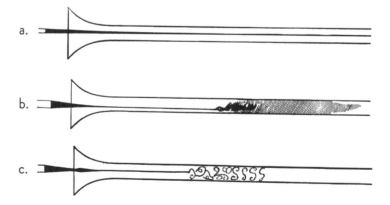

The types of flow observed by Reynolds. (a.) At low flow rate, he observed a steady line of dye in the center of the tube. (b.) As he increased the flow rate, the dye filled the tube in what he called a "colored cloud" of "uniform tint." (c.) When he strobed the tube with a spark, he noted that the "uniform tint" was eddies of liquid.

As the flow rate increased, the color band remained steady until about two feet from the iron pipe opening. At this point, the band expanded, mixed with the water, then filled the tank with a colored cloud of a tint paler than the dye because it was diluted. This uniform tint was, as Reynolds learned when he flashed the tube with a spark, a series of eddies, curling liquid—"a wriggling sinuous action," he said.[24] These eddies sprang into existence suddenly, and they would disappear and reappear as the water continually crossed the pipe from one side to the other. He repeated his experiment with smaller and larger tubes and found that the smaller the tube, the faster the water must flow to create turbulent flow.

With this experiment, Reynolds highlighted key properties of the transition from laminar to turbulent flow. First, the transition occurs abruptly: the eddies do not start small and grow and grow but appear suddenly and fully formed. Second, below a particular flow rate, no turbulence occurrs. Third, there is an upper limit to the flow rate above which smooth flow cannot be sustained.

His interpretation of these phenomena drew on his unusual intellect: he always reasoned about phenomena via analogy, and although his analogies in the long run were superficial, his insights still guide engineers. He thought of water as particles—not atoms but particles of a finite size—whose motions were like the those of members of a military troop. A marching troop could stay in line or break into a scramble with soldiers going in every direction. He pictured the dye in his water as a formation of soldiers, where the line of men in the center wore brightly colored uniforms. In "smoothly flowing" troops, that line stays steady, moving toward its destination. When order breaks down, the brightly colored soldiers, like turbulent flow, will veer all over the formation.[25] The orderliness of marching troops depends, he said, on three characteristics: discipline, speed, and size. A poorly disciplined troop, a fast-moving troop, or a large troop will struggle to stay in formation, although with discipline, a fast or large troop could stay in order. What is true for troops "is exactly true for water," which, he said, "will move in a perfectly direct disciplined manner under some circumstances, while under others it becomes a mass of eddies and cross streams, which may well be likened to the motion of a whirling, struggling mob where each individual particle is obstructing the others."

The "discipline" for a fluid is its viscosity, its resistance to flow. If Reynolds were to run maple syrup through his apparatus, it would require a much greater velocity for the smooth, laminar flow to become turbulent. The speed of the troop corresponds to the flow rate of the fluid and the size to the diameter of the pipe. In a larger pipe, the transition from laminar to turbulent flow occurs at a lower flow rate than for a smaller pipe.

Reynolds gathered these observations into a single relationship:

$$\frac{\text{flow rate} \times \text{pipe diameter}}{\text{viscosity}}$$

For his smooth-walled pipe, Reynolds demonstrated that when this combination of variables was below 2,000, the flow was laminar, and when well above 4,000, the flow was turbulent; for values between, the flow was a mixture of both types. This simple relationship among a few properties of a fluid informed the designs of engineers for over a century.

Reynolds's work exemplifies one of the main methods engineers use to handle uncertainty. Instead of searching for a fundamental, perhaps molecular-level explanation of why the transition occurs—something that is still not yet fully understood—he described the flow *phenomenologically*, meaning he described the observable elements of flow but did not reveal its fundamental nature. Reynolds's work is classic engineering: he lumped the details of a complex process together into an abstracted representation that's only as complex as necessary to achieve his goals. A fluid's flow need only be described by its viscosity, flow rate, and the dimensions of the channel it's flowing in. With these, an engineer can avoid the uncertain regions where fluid might be laminar or turbulent and stay in the regions where it is smooth or violent as is appropriate—all without a fundamental understanding of why the flow is one way or another.

Reynolds's approach, a scientist might well complain, hides all interesting and relevant aspects and doesn't solve the intractable, nightmare problem of turbulence. This line of thinking underlines the striking difference between science and engineering touched on earlier: the scientific method strives to reveal truths about the universe, while the engineering method seeks solutions to real-world problems. In no way did Reynolds reveal, as he hoped, the mysteries of fluid motion, but he did help generations of engineers change the world.

In truth, most engineered products and systems in our lives are poorly understood yet robust in operation—we can't fully explain why they work, but we have an amazing ability to make them work. What drives this is an engineer's sense of urgency, yet another characteristic that differentiates engineering from science. Because engineers respond to real-world needs—we want our buildings, cars, and medicines now!—they often work with an attitude of "we need to figure this out now" that contrasts sharply with science's leisurely investigation of the world. Nowhere is this more apparent and stunning than in the networks connecting our computers.

No one knows what happens with the billions of bytes of information that course through the networks between office computers or in the manifold devices in our homes that share an internet connection, which means that at any moment, the information in our technology is in danger of destroying itself. The information-carrying signals zipping through these networks are always at risk of interfering with one another to create only a garbled, unintelligible mess, like two people trying to talk over each other. To computer engineers, this is called a "collusion." Yet byte after byte, trillions a second, effortlessly pass through a local network, out of offices and homes, to share information around the world without distortion and disruption because an engineer devised a clever way to overcome the uncertainty about what exactly is going on in those cables. The impetus for solving this problem was a most ungainly invention: a modified Xerox-brand photocopier—its top ripped off and replaced with black-tubed lasers, mirrors, and a mass of wires.

In the early 1970s, Bob Metcalfe was puzzled by a problem: how to send documents to this first laser printer from a hundred connected computers without signals colliding and rendering the network useless. At Xerox's Palo Alto Research Center (PARC), the company's research lab in the 1970s, he saw the future of computing about ten years before it happened. Surrounding Metcalfe were a slew of personal computers, called the Altos, which used a graphical user interface like a Mac or Windows computer today, equipped with an early version of the mouse and all hooked to Xerox's latest gadget, this laser printer. Metcalfe's task, to share the printer with a hundred computers—to network it, in today's parlance—was not a pressing problem outside PARC. Before the PC, a computer cost about a quarter million dollars, so few companies had more than one, and the home computer was a fantasy.

Around this time, Metcalfe had couch surfed at a friend's house in Washington, DC. Jet-lagged and unable to sleep, he grabbed from a shelf a thick book of technical proceedings. Unfortunately for his plan to adjust to an East Coast sleep schedule, one subject captured his interest: the ALOHA System—a radio network for communication developed to transmit signals among the Hawaiian Islands, which was tackling a problem uncannily similar to Metcalfe's. If two different stations on the islands were to transmit a message at the same time, their signals would collide and not be received by whoever was supposed to get them. But there was, of course, no way for one station to know when another was about to transmit. The method used by the ALOHA System, now often recalled as ALOHAnet, was simple: any station could transmit a message when it liked, then wait for an acknowledgment that its signal had been received. If no acknowledgment returned in a short time, the sending station assumed that another station had also transmitted at the same time and that the signals had collided. The sending station would wait for an entirely arbitrarily chosen amount of time before sending again, and as long as the amount of time they waited was different from the random amount of time the station with which they had collided waited, their next attempt would go through. Even if they did happen to transmit at the same time again, all they had to do was wait a bit longer.

To Metcalfe, the solution was elegant. "It achieved," he wrote, "reliability through simplicity."[26] He noted that as traffic increased on the ALOHA System, the collision rate would rapidly increase as well, so he supercharged the basic method of ALOHAnet in order to network multiple computers to a single printer. He devised a way for a computer that was about to communicate with the printer to listen for signals from other computers on the line. The computer sending a signal to the printer could then detect whether its signal had collided. When this happened, each computer that had sent a colliding signal listened for silence on the line and then resent their data, but only after they waited a random amount of time. If they both resent right away, the signals would again collide and become stuck in an infinite loop of destroyed data. In this method, there was no need for certainty about what was going on within the network. The uncertainty of randomness was built into the solution: just pause for a bit and try again. It seems that from uncertainty stacked on uncertainty, chaos would result, and in fact a joke at the time was that Metcalfe's method "works in practice but not in theory."

But to become the industry standard, Metcalfe had to battle a rival method promoted by the titan of the computer industry, International Business Machines. IBM was bothered by the probabilistic nature of Metcalfe's solution. Their biggest clients were large enterprises, financial monoliths described by an IBM engineer as "pretty traditional institutions."[27] For these conservative organizations, the preferred approach to networking was captured by an IBM engineer: "We have to know it is going to work; our customers are going to shoot us if it doesn't work."[28] The best networking for IBM, then, was one developed in the late 1960s by a Swedish inventor, Olof Söderblom, for a large Swedish bank. They needed to network twenty-five hundred teller and office terminals at five hundred branches scattered across twelve hundred miles. Unlike Metcalfe's, Söderblom's solution removed *all* uncertainty about whether a cable was available: Söderblom's approach embodied totalitarianism in its approach to uncertainty, in contrast to Metcalfe's networking by laissez-faire and near chaos.

Söderblom linked computers along a ring-shaped cable. A "token"—a bit of information that indicates whether a signal was on the cable—circled the ring. This token worked something like a "talking stick" used in a group to designate turns to speak. When a computer had something to send, it waited until the token came by, read it to see whether the ring was being used and, if not, it grabbed the token and sent its information, confident that no other messages would be allowed on the ring. Once the signal arrived, the token was released, and other computers could use the cable. This token removed any uncertainty that other signals were on the cable. This "token ring" system was licensed by IBM to sell to large companies.

In the early 1980s, Metcalfe's methods and token ring networks battled to dominate the market, but token ring had three key advantages. First was the raw financial might of IBM. In the late 1970s, it was the largest company of the S&P 500, making up 7.2 percent of the index at one point. Its size and might rivaled that of Exxon, AT&T, and General Motors. IBM's dominance as *the* computer company was unquestioned: it often controlled 60 to 80 percent of national computer markets across the globe. Second, the token ring protocol was easier for an IT department, since IBM typically built the ring in a single room. This meant that to fix the network, a technician needed only go to a single room to troubleshoot and repair. This infrastructure also matched the existing topology of telephone wiring in a commercial building, which led to many token ring installations in the central telephone rooms. In contrast, Metcalfe's method used an unrooted tree—a mass of cables connected, like offshoots, to one central long cable—which was effective and practical in small installations but became a nightmare in large businesses. And third, token ring used a twisted pair wire cable, much more flexible than Metcalfe's coaxial cable—what we use today for cable TV. So early in the battle for superiority, token ring had an edge, but it did not have a champion with the vision and charisma of Bob Metcalfe.

The six-foot-one Metcalfe cut a dashing figure with his thick head of hair, bushy red beard, and wing tips. "I loved wing tips," he said, "because I had been to the Sloan School of Management as sort of a semi hippie, a right wing hippie."[29] His commanding presence convinced many companies to take a chance with "Ethernet," which Metcalfe, in a savvy marketing move, had named this protocol. He coined the name from the luminiferous ether, thought in the nineteenth century to be necessary for light transmission. A PARC engineer recalled that "everybody latched on to it immediately. It was very, very obvious from day one, that that was the kind of name that really stuck."[30]

Metcalfe's natural charisma was enhanced by a clear view of the future, not open to his competitors, of the dawning age of the personal computer. This knowledge, one of his colleagues reflected years later, endowed Metcalfe with apparently messianic powers. At Xerox PARC, Metcalfe saw the rise of the personal computer, while IBM thought only of the large, centralized computers used by their enterprise customers—IBM didn't even bother to put a networking port on the personal computer it sold. With this vision, Metcalfe knit together a set of nimble and determined companies to sell Ethernet to manufacturers of PCs. He pushed Ethernet as an open protocol that encouraged all to innovate collectively to improve it—a full embrace of the engineering mindset that seeks all input, all possible heuristics, that might find ways to work around uncertainty. In contrast, IBM never developed a set of such supporting companies. Even though token ring was an open standard, IBM didn't play friendly. They licensed Texas Instruments to manufacture token ring chips but forced the company to lag behind: when IBM developed improved chips, Texas Instruments had to wait until IBM manufactured and sold the new chip before they could do so. And when a company wanted to sell a computer system with token ring connectivity, Olof Söderblom and his business partners demanded exorbitant royalties. The final nail in the token ring protocol's coffin was its rigid, deterministic nature, sitting within the fastest-changing technical industry in history. It couldn't scale along with the adaptive Ethernet as computers grew more powerful and customers expected them to be faster. Ethernet eventually passed token ring in technical superiority, with speeds soon outstripping it by many orders of magnitude.

While the methods of Reynolds and Metcalfe are ingenious

ways to overcome uncertainty, they leave open the idea that working around uncertainty is only a stopgap measure until science can help us proceed with the "real" solutions. We might think that, extrapolating from the rapid advance of scientific inquiry in the last two hundred years, we will reach a time when many things are so deeply understood that no approximations are needed. This hope that science will subsume all, that it will bring perfect exactness and clarity to engineering practice, misses the fact that the reason engineering exists is to work at the limits of scientific understanding and reach beyond codified knowledge. Scientific breakthroughs only push out the boundary between the certain and uncertain, the boundary where engineers work.

In the age of modern science, beginning in the early nineteenth century, the pattern has been that a scientist observed a phenomenon, an engineer used it to create something, then much later, that phenomenon led to a new scientific understanding. This pattern has existed since what is known as the "Great Victorian Inheritance," a period of scientific and engineering growth that laid the foundations for our twentieth- and twenty-first-century technologies.[31] Scientific observations of the early to middle nineteenth century in chemistry, medicine, electromagnetics, and quantum physics eventually led to a new scientific understanding of the world, yet before this new knowledge crystallized, engineers used these phenomena to change the world. Chemists synthesized long, rubbery molecules, and as they puzzled about their nature, Hilaire Bernigaud spun miles and miles of "Chardonnet silk." Scientists discovered that a current passed through cables could control a magnetic needle, a baffling phenomenon intractable to the theories of the time, yet engineers built vast telegraphic systems under the ocean. Scientists identified microorganisms as a source of disease but lacked enough understanding to design a "magic bullet" cure that would target these microorganisms, yet in 1928, Alexander Fleming, using the engineering method, discovered penicillin. It was manufactured on a large scale by the 1940s, forty years before its mechanism of antibiotic action was understood. And physicists observed the photoelectric effect, a quantum phenomenon that defied any classical description, yet an engineer used the photoconductivity of glassy selenium to create a photocopier in 1938— thirty years before scientists fully understood the phenomenon.

These examples, though, hail from the scientific discoveries of the late nineteenth and early twentieth centuries, so surely the stunning advances in scientific understanding and techniques in this century will remove the uncertainty that has plagued engineers in the past. Yet nothing of the sort happens, because as scientific knowledge advances, engineers step beyond that knowledge. Consider an example from the premier science of the last fifty years, molecular biology. Deciphering the code of life embedded in DNA opened a deep and rich mine of knowledge about how organisms work. An American chemical engineer pushed past the boundaries of this deeper scientific understanding to create enzymes that reduce the environmental cost of producing our fuels, pharmaceuticals, and chemicals.

Enzymes are similar to catalysts—a substance that enables a chemical reaction but doesn't itself change—but where catalysts are simple, inorganic molecules, enzymes are created by and enable the creation of life. In baker's yeast (the Greek word for yeast being the origin of the word *enzyme*), an enzyme breaks down sugars to create the carbon dioxide that causes the dough to rise. When we inhale air, an enzyme catalyzes the breakdown of oxygen into the waste products water and carbon dioxide that we exhale. These are only two of the thousands of chemical reactions catalyzed by enzymes, reactions by which cells feed, grow, reproduce, excrete waste, move, and communicate with other cells.

Enzymes themselves are complex protein molecules. Typically five hundred amino acids, of which there are twenty types, are linked to form an enzyme, which means there are 20^{500} possible combinations of amino acids that can create an enzyme—a mind-bogglingly large number, well beyond the number of atoms in the universe. Finding new and useful combinations among the astronomical possibilities baffled scientists but did not prevent engineer Frances Arnold from creating enzymes as solutions to specific problems.

She drew insight from a short story. "I'll never forget," she said, "reading Jorge Luis Borges's short story 'The Library of Babel' when I was working in Madrid in the summer of 1976."[32] In 1941, Borges described the astronomical scale of uncertainty inherent in our collective knowledge. His story "The Library of Babel" imagines an infinite library of hexagonal rooms, each containing shelves of books that are full of randomly assorted letters and punctuation, such that within the library is every possible combination of words that can create a coherent thought, including perfect descriptions and explanations of the world and even predictions of the future. Although the librarians who roamed the bookshelves hoped that some text might reveal "the fundamental mysteries of mankind," they found for every rational thought in the books, there were "leagues of senseless cacophony, verbal nonsense, and incoherency."[33]

To Arnold, the books of letters were an analog to the inconceivably large number of possible arrangements of the amino acids in an enzyme. "This code of life," she said, "is a symphony, guiding intricate and beautiful parts performed by an untold number of players and instruments. Maybe we can cut and paste pieces from nature's compositions, but we do not know how to write the bars for a single enzymic passage." She wondered whether she had "joined the Babel librarians in a hopeless search for these magnificent but extremely rare books." But in quick succession, she realized three things that could order her enzyme "library" and lead to new, powerful enzymes tailored to work outside the human body.

First, unlike the librarians of Babel, who had no idea where to find books that contained anything other than gibberish, Arnold knew that useful enzymes already existed. "They are everywhere," she said, "and can literally be scraped from the bottom of my shoe" or "captured from the air I breathe."[34] These, she reasoned, were the useful books in her enzyme library. Second, nature evolved these useful enzymes with small changes over billions of years, by the replacement of single amino acids one at a time. These enzymes could lead her step by step to other useful enzymes. And third, she realized that nature's own method was how to create industrially useful, robust enzymes: evolution. "Nature," said Arnold, "by far the best engineer of all time, invented life that has flourished for billions of years under an astonishing range of conditions."[35] By directed evolution, Arnold thought she could harness the nimble, adaptive qualities of enzymes and direct them along paths nature left unexplored.

Her idea of "directed evolution" of enzymes met resistance from scientists. Those who wanted to *understand* proteins were "aghast," crying, "That's not science!"[36] She responded by explaining, "I'm an engineer," noting her goal was the engineer's guiding principle of "getting useful results quickly."[37] Others told her it could not work because nature had already optimized enzymes over billions of years, that any and all useful combinations of amino acids would have been discovered by the immense power of that process, but she realized this reasoning was faulty: nature had explored only a tiny fraction of life's possible molecules, "precisely because nature did not ask for these behaviors."[38] She thought the engineering method would work. With that key insight, Arnold entered the territory most fruitful for an engineer: working on the margins of solvable problems, even those unexplored by nature.

Her first engineered enzyme was synthetically evolved from a member of a group of enzymes that gives us the ability to digest milk.[39] Originally evolved to allow us to feed on our mother's milk as infants, most modern humans have adapted to keep making the enzyme lactase into adulthood in order to consume dairy products. In the water-rich liquids of the small intestine, the enzyme works astonishingly well, but when Arnold put it in an organic solvent called dimethylformamide—think of paint stripper, and you have a pretty good idea of what this chemical is like—the enzyme no longer, of course, "digested" milk proteins. Arnold then simulated evolution by creating mutated versions of the enzyme, hoping at most to have changed an amino acid or two, and tested the mutations. This time, most failed to digest the milk protein, but a few managed to succeed at least partially. She selected the best new enzyme, created mutated versions, and tested again. After ten rounds of mutations and selection in increasingly higher concentrations of the solvent, she engineered an enzyme that worked in a harsh chemical environment almost as well as the original did in water. At first, she was surprised how few mutation cycles were required, then realized how nimble nature's evolutionary apparatus was: "Genes, like ideas, move around. Good ones get picked up quickly and spread even faster. If you do not believe it, just try to keep ahead of drug resistance."[40] This newly "evolved" enzyme proved her idea of directed evolution as the way to engineer useful enzymes.

In a sense, biological evolution and the variety of species form the perfect example of engineering despite uncertainty and scientific ignorance. Evolution as a process has no knowledge of how enzymes work or even what they are; it doesn't know what chemical compounds are available to it or what it has created with them. For evolution, the possibilities are endless. It has absolutely no knowledge of anything whatsoever, but uncertainty is no object because it is, by its essential nature, an imperative within the physics and chemistry of life to solve problems. Where the sheer volume of uncertainty had daunted researchers, Arnold had the intuition to apply nature's own engineering method. Paradoxically, this uncertainty, which from the viewpoint of science is a deficiency, was to Arnold, as it is for all engineers, freeing: it lets the imagination run riot. The results speak for themselves. In the decades since her work, enzymes engineered by directed evolution have diagnosed and treated disease, reduced farm waste, improved textiles, synthesized industrial and pharmaceutical chemicals, and enhanced laundry detergents—that enzyme that breaks up milk protein also breaks up stains!

"A wonderful feature of engineering by evolution," Arnold said when accepting the Nobel Prize, "is that solutions come first; an understanding of the solutions may or may not come later."[41] That deep understanding of enzymes has yet to arrive: "even today," she noted, "we struggle to explain" how her evolved enzymes work.[42] This is a clear reminder that as our knowledge about the universe expands, an engineer will always be out front, working in the penumbra of understanding. Because advances don't remove uncertainty. They simply move the borderline between certainty and uncertainty—the perfect space for an engineer to work.

4

RESOURCES

From Mesopotamia to the Islamic Empire
to Space—How Materials, Energy, and
Knowledge Shape Everything

In the seventeenth century BCE, Sidqum-Lanasi, a senior official in the city-state of Carchemish, on today's Turkey-Syria border, faced a gnarly problem: how to ship eighteen thousand liters of wine he'd sold in a side hustle to the king of Mari, a neighboring city-state in politically fragmented Mesopotamia. For Sidqum-Lanasi, the deal promised large profits—a bottle of wine sold in Mari for three times what it cost him to buy in the agriculture-friendly climate of Carchemish, which produced vineyards, forests, and fields of wheat. The king of Mari and his people lived on the salty, barren soil of the Syrian desert, baked by nine-month-long summers with highs of 118°F, freezing winters, and a spring filled with dangerous sandstorms. But Mari's curse was a bountiful opportunity for Sidqum-Lanasi only if he could transport the wine 250 miles to Mari with two life-threatening options to get there:

a pack-animal caravan down paths patrolled by bandits or a ride down the Euphrates River.

Zimri-Lim, the king of Mari, sustained his kingdom through his one geographical advantage: trade up and down the Euphrates had to go through him, and he was sure to extract hefty taxes from the trade of hardwoods, exotic foods and animals, precious stones, and tin for manufacturing bronze. With vast wealth, Zimri-Lim built a 260-room palace of sunbaked mud bricks; its sixteen-foot-tall walls enclosed nearly three hundred thousand square feet. Inside the walls, Zimri-Lin created a verdant paradise, unimaginable in the desolate wasteland outside the palace walls. A canal from the Euphrates irrigated a lush royal garden filled with cedar, cypress, date palms, olive trees, and the aromatics loved by Mesopotamians—myrtle shrubs and pomegranate trees that, as the royal gardener noted, "enrich the breezes."[1] These gardens charmed the ambassadors, diplomats, and dignitaries who traveled the Euphrates as, from the garden, Zimri-Lim's guests entered the palace through a doorway outlined in limestone but painted to imitate marble. Everything displayed his power and wealth, an impression reinforced by a wall painting of the king with a dagger in his belt and two goats in a heraldic pose, nibbling on a sacred tree to remind viewers that the king was an intermediary to the gods.

Once guests were inside the temple-like throne room, an unearthly chill sent shivers down their spines. Large pools of water built within the palace walls were designed to freeze during the cold, cloudless, desert nights, then cool the air that flowed through the palace during the sweltering day.[2] Once guests had assembled, the chilly, somber throne room burst into festivity as food flowed in from adjacent rooms—pickled grasshopper from Assyria, fish roe from the Balikh River, deer meat from the nearby mountains, and, in springtime, truffles. To wash this down, Zimri-Lim served date and barley beers—light, dark, amber, and even filtered—although most guests hoped for sweet red wine, served chilled with ice, a treat too expensive for all except a king. To protect the wine, Zimri-Lim ordered each jarful stamped shut with a royal seal to prevent pilfering, then stored in a locked cellar with a key he kept personally.

To buy this wine from Sidqum-Lanasi, Zimri-Lim sent $2\frac{1}{6}$ pounds of silver, 130 shekels in the currency of the day, with which Sidqum-Lanasi had to buy six hundred red clay jars, the wine itself, and the means to ship it.[3] As he fretted about the problems of transportation, he begged Zimri-Lim to be patient: "I hope my lord will not say anything in his heart about the delays. The land is stirred up and routes are cut." He later wrote, "In last night's attacks my sheep were carried off. My lord knows I will not be able to subsist without sheep." The clay jars cost him ten shekels, the wine ninety shekels—"wine has become expensive," he wrote to Zimri-Lim—which left thirty shekels to spend on getting the wine to his powerful client.

With the land "stirred up" by raiding bandits, a journey by boat was the only remaining option. But the Euphrates was a river full of hazards, the most extreme being the rapids that smashed boats to pieces against the river's rocky bottom. And even if a boat and its cargo survived, going by river was a one-way trip. The strong current, not to mention the rapids, kept boats from returning upstream to Carchemish.

To deliver the king's wine, Sidqum-Lanasi had to rely on a unique solution devised by traveling Mesopotamian merchants and dictated by both an abundance and lack of resources. The upper Euphrates had agriculture, timber, animal husbandry, everything that defined the Fertile Crescent of the Middle East. Yet the dangers of the Euphrates made the kinds of rigid wooden boats they could build for a trip like this with quality timber completely useless—they would be torn apart by the first run of rapids—and the impossibility of returning the boat upriver made it an enormous waste of time and material. The fatal missing resource required that the boat be cheap and disposable, a raft purpose-built for the task: a *kelek*.

A kelek, a raft used on the Euphrates. These rafts were formed from large tree trunks, laced together with twigs, and protected by air bladders made of sewn-together goatskins. A raft was typically fifty feet square and could haul several tons.

The foundation of these fifty-foot-square wooden rafts was formed from large tree trunks, laced together with twigs from a small Eurasian willow used in basketwork, then covered with smaller tree trunks to create a platform, also bound by twigs. Air bladders, fashioned from about one thousand goatskins, were attached to the bottom of the kelek, inflated by blowing through a reed tube as water was poured over the bag to keep the skin from drying and leaking, then tied at the neck with a thin twig. These bladders protected the raft from the rocky bottom of the Euphrates. In slow regions of the river, the raft's two front oarsmen propelled the raft; in rapids, the rear rudder man strained to steer the raft as it shook and creaked, while the two oarsmen scrutinized the river, alert for shallows, ready to dip their oars to slow the raft as it scraped across jagged stones. These stones deflated the bladders, which the pilots reinflated or replaced in calmer waters. As they did this, the smell of roasted gazelle, deer, or waterfowl wafted from the raft as dinner cooked with bread cakes, eggs, and preserved dates as sides. In these quiet moments, the gentle ripple of the water was punctuated by the braying of donkeys—a resource as essential as the wood in delivering the wine. Once the rafts docked at Mari, the crew unloaded the wine and donkeys, removed the goatskin air bladders, then with a flick of a knife sliced the thin twigs binding the rafts' foundations. The crew sold the wood, precious in resource-poor Mari, dried the skins, packed them on the donkeys, then returned by road north to Carchemish.

This solution of a kelek to transport the wine captures the subtlety in the phrase *using limited resources* in our definition of the engineering method, a phrase that at first seemed banal (who ever works with *un*limited resources?) and, at its simplest, a truism. If an engineer lived in the Bronze Age, then they used, of course, bronze! This simple but not incorrect gloss misses the nuance of how engineers weave available resources in subtle and unexpected ways and how often the *absence* of a resource sets the final look and feel of an object. In the kelek, we see how the available resources influenced the raft's design: wood plentiful in the north, scarce and valuable in the south; skins time-consuming to prepare but reusable and light to transport; donkeys used to return quickly to the north to build another raft; and the fundamental nature of a raft that can be easily disassembled, sold, and repackaged dictated by the natural flow of the river that created a quick, if perilous, journey toward the Persian Gulf but flatly denied any trip back. This mix of resources might strike us today as quaint, but this ancient supply chain is as complex and intricate as any mass-produced engineering project today. "Limited resources" have exposed and will always expose the details of how engineers execute their designs and overcome uncertainty to respond to urgency. Each of us has seen in our lifetimes products that evolved to better and better materials or that work better and better every decade—the reliability of a car today is stunning compared to what it was fifty years ago—yet the lack of the ultimate materials or perfect performance didn't prevent these products from being designed or brought to market. In every case, an engineer responded with urgency to the needs of the time as well as possible.

The story of the kelek highlights material resources, and although the ingenuity and skill of an engineer in marshaling materials is rarely appreciated, the idea is familiar. The moment we touch an object, we can tell whether it is plastic, metal, ceramic, or wood. Each texture is immediately imaginable. But the role of another resource is often hidden from view and is intimately, intricately, and inseparably intertwined with engineering design: energy.

In the West, we often trace the roots of technology backward from the Industrial Revolution to the Enlightenment to the "Dark Ages," when a few monks squirreled away scientific and cultural knowledge, keeping alive the flame of classical wisdom from the spark of all learning lit by Greece and Rome. But while the Europeans were hiding books in monasteries, the emergent intellectual traditions of the Islamic world enabled it to stretch from as

far north as the snow-capped Pyrenees, around the southern shore
of the Mediterranean, to the monsoon-swept city of Calicut, today
Kozhikode, a mere 250 miles from the southern tip of India. The
early Islamic caliphates had the technology to create large cities—
Baghdad's population was greater than 1.5 million in the tenth
century, four hundred years before the population of Paris topped
100,000—and to feed a vast, geographically diverse population.
They harnessed the power of wind and water to drive mills that
ground grain and powered mechanisms that husked rice, pulped
sugarcane, processed cloth, and crushed metallic ores.[4]

Water and wind drove the mills that fed the body, but the
work of Islamic engineers was also to nourish the soul. One of the
five "Pillars of Islam," *salat*, mandates that faithful Muslims pray
five times a day within precise intervals. At sunset, the beginning
of the Muslim day, the evening prayer can begin but must occur no
later than nightfall. The night prayer can begin any time between
nightfall and daybreak. The morning prayer occurs between daybreak and sunrise. And finally the midday prayer begins when the
sun has crossed the meridian and lasts until the start of the after-
noon prayer. Islamic engineers' earliest timekeeping methods were
detailed tables that reported time as a function of the height of
the sun in the sky, but these tables worked for only a small region
around a particular latitude, with the first tables developed for
large cities like Cairo and Damascus.[5] Although the tables worked
well enough for the public, faithful tinkerers and their patrons were
interested in more precise methods of timekeeping that could tell
a Muslim exactly the times of day they should dedicate to prayer.
The clocks developed by Islamic engineers place them in the front
ranks as virtuosos in the controlled release of energy to drive their
mechanisms.

The greatest of these was Abū al-Izz Ismaīl ibn al-Razzāz

al-Jazarī (or al-Jazarī for short), the twelfth-century chief engineer for one of the many minor princes in the Islamic world. At age sixty-five, he poured a lifetime of experimentation into *The Book of Knowledge of Ingenious Mechanical Devices*, in which he distilled the fruits of his peaceful life spent under the protection of an Artuqid dynasty prince, a vassal of the powerful Saladin who had vanquished Richard the Lionheart during the Third Crusade. This stability—only, though, if he stayed put, since to travel risked encounters with marauding bands of Kurds—enabled him to do what he liked best. As he wrote in his *Book of Knowledge*, that was to "contemplate in isolation" driven by a thirst to invent described as "engrossed in diligence" and consumed by "strength in passion."[6]

The book detailed devices far ahead of their time in fragmented and warring Europe: a pitcher with a mechanism that mixes cold and hot water to create a stream of warm water at a fixed temperature, fountains that throw water into the air, a combination lock, and machines that draw water and facilitate irrigation. It's filled with prescient descriptions of what is now mundane engineering practice: the lamination of timber to minimize warping, today's plywood; the balancing of wheels, now done routinely on every car; and wooden templates and paper models to design objects, the blueprints of yesterday and now computer drawings. But as an engineer in a noble's court, al-Jazarī had both the opportunity and mandate to turn engineering into art that entertained and inspired wonder. Arguably the creator of the first "robots," he designed automata that poured drinks, played music, and offered towels to his prince's guests. All his designs revealed the precision necessary to be a good engineer—the calibration of water's flow rate with orifice size, or the creation of watertight valves by grinding the valve's seats and plugs with emery powder. This care, precision, creativity, and religious duty led to the pinnacle of al-Jazarī's engineering: clocks. *The Book of Knowledge* lovingly details ten types of clocks, all featuring automata—birds that drop pellets on cymbals, rotating zodiac circles, musicians who strike drums and play trumpets—that decorate miniature castles, boats, and elephants. But the operation of his most clever clock, called "the candle clock of the scribe," was intertwined with its energy resource.

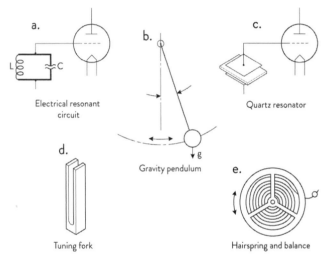

Resonance is the tendency of a system to oscillate with larger amplitudes at some frequencies than at others; these points are called a system's resonant frequencies. All resonators, if moved from rest, transform their stored energy back and forth from potential to kinetic at a rate depending on the mass and stiffness of the spring or pendulum or equivalent electrical properties. At each oscillation, that is, the change from potential to kinetic energy, the resonators lose a small portion of their energy to internal friction and so eventually decay. There are many ways to create resonance, all of which can be used to tell time. Shown in this figure are (a.) an electrical circuit with an inductor and capacitor; (b.) a pendulum driven by gravity; (c.) a quartz resonator; (d.) a tuning fork made of steel or quartz; and (e.) a hairspring with a balance as used in early twentieth-century watches.

The history of timekeeping devices is of mechanisms that regulate the release of energy, dissipating it in uniform increments.[7] The first clocks drew their energy from gravity: the flow of water into a basin or the trickle of sand grains in an hourglass. This era of *continuous flow* lasted until the fourteenth century, when a device called an escapement—the most common form was a version called a verge and foliot—was created to convert the energy of a falling weight into vibratory motion. The weight was attached by a cord to the axle of a wheel, and as the falling weight spun the wheel, the escapement momentarily halted the rotation of the wheel and the motion of the weight. The rate of these stops and starts, these oscillations, regulated the clock. Although clocks proliferated in this era of "aperiodic control"—from large clocks for towers to small clocks for homes—these mechanical clocks were poor timekeepers. The reported time varied as much as two hours per day, so clocks were built with only hour hands.

In the next epoch of timekeeping, the escapement was still central, but the energy dissipation it controlled came from sources more easily regulated than a falling weight. The final step in timekeeping is the "resonant control" era, which began in 1656 when astronomer and mathematician Christiaan Huygens devised the first pendulum clock. The pendulum stored energy, then, as it swung, dissipated it at a constant rate. The advantage of a pendulum over a falling weight is that much of the energy stays stored in the pendulum; not all of it is removed with each swing. This allows the pendulum to swing freely and keep better time than earlier mechanisms. This central idea evolved into clocks and watches powered by the unwinding of a coiled spring, then the vibration of a quartz tuning fork, and finally atomic clocks based on quantum mechanics. None of these methods, of course, were available to al-Jazari, who lived in the era of continuous flow timekeeping, yet in his candle clock of the scribe, he devised a most ingenious mechanism that used neither water nor sand.

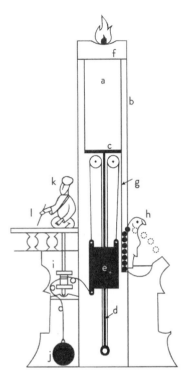

This clock appeared at first to be only a decorated candle: a brightly polished brass tube, which sits in a shallow brass dish, with a flickering flame at the top. Mounted on the tube's sides were two brightly painted figures: a falcon on one side and, sitting on a shelf suspended from the brass tube, a seated scribe wearing a turban and a flowing robe and holding a long pen in his right hand pointed toward the shelf. His head bent forward, he concentrates on a dial divided into fourteen and a half major divisions, each representing an hour.

A falling weight powers Islamic engineer al-Jazari's "candle clock of the scribe." The rate at which this weight falls is controlled by a burning candle. A properly sized candle (a) is placed inside the brass tube (b) of the clock. The candle rests on a movable metal plate (c). This plate is affixed to a long rod (d) that passes through the center of a hollow weight (e) A. The weight is connected to the long rod by cords that pass over pulleys, through the weight, and are tied to the bottom of the rod. As the candle is inserted, it raises the weight; then the candle is covered with a cap (f) with a small hole for the candle's flame. The weight presses the candle against this cap. As the candle burns, the plate (c) rises and the weight drops. Attached to the bottom of the plate is a channel (g) containing metal balls. As the plate rises, the metal balls drop through the mouth of the decorative falcon (h). A cord at the bottom of the weight turns a spindle (i) — the cord held taut by weight (j) — which rotates the scribe (k). As the scribe rotates, his pen (l) traces out the minutes on a dial. In this drawing, adapted from a sketch in al-Jazari's *Book of Knowledge of Ingenious Mechanical Devices*, the size of the spindle underneath the scribe and of the balls is schematic: if one were to build this clock, the spindle would be sized so that the scribe completely rotates once every fifteen hours and the balls sized so that one ball drops from the falcon's mouth every hour. Typically, the candle used in the clock was sized to burn for fourteen and a half hours.

As the scribe slowly rotates, the tip of his pen, which hovers above the dial, indicates the hour or fraction of an hour divided into four-minute intervals. When the pen points toward one of the hour divisions, a brass ball, perhaps a quarter inch in diameter, rolls from the falcon's mouth, dropping into the brass dish with a distinctive plunk.

All this action is driven by a dropping weight inside the brass tube. As the weight drops, it tugs on a silk thread that rotates the scribe by turning a spindle hidden beneath him. Another thread from the weight passes over a pulley and raises a rack of fourteen brass balls, such that once an hour, a ball matches an opening behind the falcon's neck and slides through its mouth. The energy, then, that powers this clock comes from gravity, the potential energy of the weight. The key engineering insight was how to *control* the release of that energy, to steadily drop the weight. Al-Jazarī ingeniously controlled the weight's descent with a candle.

Al-Jazarī cast a candle of "pure wax" that "weighed 160 dirhams" (*dirham* is the Arabic word for a Greek coin and equaled about a tenth of an ounce).[8] He then measured the time it took this candle to burn one hour, using what he called a "standard ruler." Once he knew this length, he cast another candle to be fifteen of these hour lengths long. The new candle was hidden inside the brass tube, its flame peeking out at the top. Inside the tube, the candle sat on a brass plate, connected by a pulley to the weight beneath. The weight pressed the candle into a plate on top of the brass tube, with a hole in the plate for the candle's flame. As the candle burned and, of course, shrank, the weight slowly dropped at a precise rate because al-Jazarī had sized the candle to count out the hours.

In this clock, we see the inseparability of engineering design and the form of energy that powers it. If a cascade of falling water powered the clock, the inside would be filled with paddle wheels. If a coiled spring, then the gearing inside would be radically reconfigured. And if al-Jazarī lived today, he would scrap everything inside the clock and use a tiny electric motor to spin the scribe and tilt the falcon. Examples abound of the same design constructed of different materials: ink pens can have a plastic casing or a metallic one; the body of a car can be steel, aluminum, plastic, or fiberglass. But a consumer's brain is not wired to as easily understand how a technology's design would change on account of the energy source. We intuitively think that building an electric or hydrogen-powered car is a matter of switching out the engine, with perhaps a few details altered to fit the bill. But the energy source is so central, so elemental, so tightly coupled to a device's operation that a change in that source most often requires a completely new design. This linkage, hidden before understanding the engineering method, is critical when technologies evolve and ignored to our detriment. Many think, for example, that to change to a hydrogen economy, we simply use the gasoline pump and fill up our slightly modified cars. Yet from source to use, the hydrogen economy would be a complete reworking of our transportation infrastructure.

Equally important as these two tangible resources, materials and energy, is the resource of knowledge. This includes the tacit engineering knowledge gained over decades of experience, that inheritance of tried-and-tested rules of thumb passed down over millennia of human history, and the knowledge unique to a person, especially what they gained from their singular experiences. An illustration of all these resources—materials, energy, and knowledge—is the rocket engines that propel satellites.

The engines, nicknamed "gentle rockets" because they lack the drama of the brilliant flame and tremendous roar of the rockets that lift the satellites into the sky, position and control the orbits of the roughly two thousand active satellites that circle the earth.[9] These satellites send television, radio, and voice signals; guide automobiles and trucks by the global positioning system; and monitor the earth, warning of cyclones, hurricanes, and typhoons. In no technology is the choice of material and energy source more exacting than in space technology. Errors are difficult to rectify, most catastrophic—resulting in inoperable satellites or orbital decay and demise. To create a successful space technology, the gentle engines of a satellite, demands a mastery of engineering from a lifetime of experience that, as for many engineers, begins in a childhood fascination with flight.

In 1934, ten-year-old Yvonne Claeys Brill ardently followed the adventures of aviators Amelia Earhart and Amy Johnson as they crossed continents with their airplanes. Their independence impressed her, their "freedom to fly" as she phrased it, and inspired her to devise a plan while riding the streetcars in her hometown of Winnipeg.[10] She and her family often rode the streetcar from their sleepy suburb to the auditorium downtown where they watched wrestling, roller-skated, competed in music festivals, square-danced, or toured an auto show, but as the streetcar neared the auditorium, Brill caught sight of the stately stone buildings of the University of Manitoba. "I just decided," she later recalled, "when I was about ten years old, that I wanted to go there." Finding herself more drawn to the machines her role models used than their adventures, Brill aspired to become an aeronautical engineer despite her parents, immigrants from Belgium, hoping she would grow up and start a small shop—her father suggested selling dresses. But Brill wanted to see the world. When she graduated from high school, she earned a scholarship to the university, where she applied to the engineering school, but she ran into the first obstacle keeping women out of the discipline. She was told by the dean that all engineers had to attend a summer camp, but the camp lacked accommodations for women and thus couldn't allow her to participate. As a compromise, she took a science degree, in the last two years concentrating on mathematics, chemistry, and physics. Yet she never lost the desire to be an engineer, nor lost her interest in aviation.

After graduating in 1945, she wanted to visit South America—perhaps inspired by Earhart's final flight that crossed the northern tip of South America. She jumped at a job offer from Douglas Aircraft in California—"California was on the way" to South America, she reasoned—and packed her meager possessions to leave for the United States without giving her parents a chance to object. In the mid-1940s, Douglas Aircraft pioneered passenger jets and revolutionary military aircraft, and by the time Brill was traveling halfway across the continent to join the team, the company had planted a foothold in exploring space. They designed rockets that would evolve into the Saturn boosters the world would see lifting the Apollo missions into space in the 1960s. Yet Brill's first engineering job was less than glamorous. She double-checked her boss's "back of the envelope" calculations for the trajectory of rockets powered by various fuels. Using an early mechanical calculator called a Marchant, she spent hours in detailed calculations, eventually going back to her slide rule because she found the calculator "clunky." Once she finished these calculations, they assigned her the task of extending the tables of thermodynamic properties. Rockets burn fuel at extraordinarily high temperatures, 5,000 K or so—more than six times the highest temperature of a household oven—but thermodynamic properties were typically only tabulated to 3,000 K. To calculate these properties, she drew on her expertise in chemistry and physics, but the work was routine.

Brill's work was typical of the work of many female pioneers in science and engineering. In crystallography, the first women in the field ground through the detailed calculations needed to locate the atoms in a crystal—dismissed by some as "intellectual knitting."[11] In astronomy, women at first classified thousands of photograph plates instead of doing the active work of observing the skies. The creators of ENIAC, one of the first electronic computers, employed women to hand calculate missile trajectories—teams left out of histories and dismissed with the collective term the "ENIAC girls."[12] All these jobs required superb mathematical and scientific skills, but the work was dismissed by their male supervisors as mere drudgery. Perhaps some of these pioneers enjoyed these exhaustive calculations, but to Brill, it was a "dead end." She saw "absolutely no future to punching these keys" of a mechanical calculator. Her only consolation was the opportunities unique to her field, including reading the top-secret German reports on the rockets planned by the Nazis at the end of the war. She marveled at the elegant designs and clever engineering solutions, which reminded her that she wanted to design *things*, not simply calculate on paper in support of someone else's engineering designs.

To work as an engineer, she left her secure job at Douglas for a start-up creating a novel engine called a ramjet—a type of engine thought to be useful on missiles. In this small company, she thrived on the practical business of getting the engine to run, measuring its performance, and especially "figuring out how you might change a design to do better"—the essence, of course, of engineering. She cut her teeth there as an engineer, then moved from California to the East Coast, where, in the early 1960s, RCA hired her as their sole propulsion engineer.

RCA was developing the first commercial communications satellites—satellites that, when launched in the early 1980s, would be used by HBO to beam their programming to local cable television operators and capture more than a million customers and 60 percent of the pay TV market.[13] Designing the onboard engines, the "gentle" engines, for these satellites was the engineering challenge Brill had dreamed of for years.

These engines are used to position the satellites once the initial launch rockets have put them in space. First, after a satellite separated from its booster rocket, it was rarely in geostationary orbit. "You might find," Brill said, "your spacecraft drifting in the wrong direction to the station you wanted it to be at," which meant you had to fire a "propulsion system pretty quickly to stop it while you still had it in your ground vision and send it the other way." The second use of these engines occurred months into the satellite deployment. Small forces perturb the satellite's orbit—the uneven gravitational force of the semispherical earth and the tug from lunar and solar gravitation—and to counter these forces, an RCA satellite fired its thrusters on a twelve-week schedule to correct its orbit. The obvious design would be two propulsion systems: one for quick, large thrusts to position the satellite in orbit after launch, and one for the smaller, scheduled thrusts to fix degradations in its orbit. But to Brill, this system would fall prey to the inescapable enemy of flight: mass. Weight is a limited and premium commodity in a satellite; a heavier satellite is more expensive to launch and uses a lesser percentage of its payload for its principal operation, as more mass requires more fuel and propulsion power to reach orbit. Two sets of orbital propulsion systems would entail two sets of piping, valves, and controls. Plus some of the fuel loaded for the "quick" propulsion might never be used, and the more accurate the initial launch, the more fuel specifically intended for correcting it would sit idle in the satellite, an unconscionable waste of weight. Every ounce of fuel used in the "quick" engines would take away from the weight of fuel that could be used for station keeping or repositioning the satellite every twelve weeks. RCA satellites needed to last five years at least (today, satellites last fifteen years), so every ounce of fuel needed to be used as efficiently as possible. The best solution would be a single system that did both tasks.

To come up with a solution, Brill drew on her comprehensive

understanding of propellants, gained from calculating those thermodynamic tables, the rocketry designs gleaned from the top-secret German reports, and her work developing the ramjet. She returned to calculating, knowing this time that what she devised, she would also create. She developed this system late at night at her kitchen table, writing on a yellow pad with one hand, adjusting her slide rule with the other.[14]

She started with chemical propulsion, like that used by the giant engines to lift rockets from launchpad to space. All rocket engines, small and large, create an explosive force by heating a chemical or mixture of chemicals to a high temperature, which creates a gas that expands and rushes through a nozzle to lift a rocket. How to raise the temperature is the important engineering bit. The Apollo mission roared into the sky on a Saturn V rocket powered by the explosion of kerosene mixed with liquid oxygen. A chemical reaction between the two components generates a high temperature and nearly instantaneous power, but it doesn't use fuel efficiently—a "fuel hog" as it was called in the industry.[15] Brill knew of a more efficient way to use fuel: instead of heating by chemical reaction, heat the fuel with electricity generated by solar panels. Brill knew of the work in research labs developing these fuel-efficient electric engines; she called them "the cat's meow," but they had a drawback: it took at least half an hour to bring them up to temperature to ignite the fuel. This delayed response was fine for the scheduled adjustments to maintain the satellite's orbit, but, as Brill knew, a quick burst from its engines would sometimes be needed to correct a significant deviation. So if a solar-powered electrical engine was to serve both purposes, it would need to use a fuel that the engine could ignite in moments rather than minutes.

To find a dual-use propellant, Brill drew on her comprehensive knowledge of rocket fuels and a rule of thumb provided to her by chemistry and physics: the hotter the rocket engine, the more efficient the fuel use. She quickly settled on hydrazine—a simple molecule expressed as N_2H_4. Pass it over a catalyst, and it will immediately and spontaneously decompose with a release of energy, quickly reaching 1,800 K. She could use less fuel by heating it even further electrically. Her encyclopedic scientific knowledge of the propellants revealed the relationships between her options, but it was the new engineering muscles she began to work that designed a solution. Brill combined all these ideas in a system that used a single tank of hydrazine and, as appropriate, fed it either through thrusters where it was heated quickly and solely by its own decomposition or to electrically heated thrusters for more efficient use when a fast, reactive thrust wasn't necessary. Thus, she created the two types of propulsion using the same system that minimized the satellite's mass, maximized fuel efficiency, and enabled a generation of satellites. In designing this engine, Brill, of course, carefully considered the material to construct it from and the substance to provide power, but her work also highlights one more resource—perhaps the ultimate resource—her deep knowledge of propulsion.

The knowledge of propulsive energy that Brill retained from her scientific education and practical experience as a number cruncher made her a uniquely valuable intellectual resource for the budding industry of spaceflight. But the fact that she was given the opportunity to use that resource to direct the future of jet propulsion design brings to mind how many similar women and other marginalized people could have added to our collective pool of knowledge if they were as lucky as Brill was to emerge from the shroud of tacit suppression. No doubt the same solution or similar ones could have been devised by others, and it's likely that others were exploring them at the same time as she was, but in all matters of engineering, time is the most valuable resource. Maximizing the breadth and depth of knowledge accomplishes the engineering method's principal goal: to shorten the amount of time it takes to devise a solution. It's a matter of luck that Brill had the opportunity to devise her solution as soon as she did, though it's very possible, almost certain, that this solution and thousands of others could have been reached even sooner had more women like and before Brill been empowered to contribute.

But Brill was ultimately among the lucky ones. She enjoyed guiding her design through prototyping and manufacturing—she was thrilled to see its gold-plated parts and shiny stainless steel tanks—but her ultimate satisfaction was to see it at work. "Nothing," she said, "was more gratifying to me than to be at a control center when the satellite finally got on orbit with all payloads functioning."[16] Her design was used on the first two hundred or so commercial satellites.

Brill's story seems moored in the mores of the late 1940s and early 1950s, yet when she died in 2012, the *New York Times* opened her obituary with "She made a mean beef stroganoff, followed her husband from job to job and took eight years off from work to raise three children."[17] After a Twitter firestorm, the *Times* replaced the first six words with "She was a brilliant rocket scientist."[18] Although she was, of course, an engineer.

5

ENGINEERING MINDSET

The Three Key Strategies That Created a Ceramic Masterpiece (and Every Other Innovation)

In mid-1774, Josiah Wedgwood opened a trapdoor in his study and descended a narrow set of brick stairs to a large, dry cellar beneath his mansion. He climbed slowly, a hollow echo in the stairwell as his wooden right leg tapped the steps. He settled into his subterranean laboratory, surrounded by nearly a hundred substances—various clays, lead for glazes, exotic materials like borax, lime (burnt and stone), pumice, iron slag, nickel, and "arsenical neutral" salt. He was pleased to be hidden from industrial spies and shielded from pestering guests—both drawn by his fame as "potter to Her Majesty" and as a pioneer of the Industrial Revolution. That day, Wedgwood intended to create a ceramic more magnificent than porcelain. This was no small feat. To do so, he must create from clay a translucent material that, although hard and brittle, looked soft, almost like ivory or vellum, and was neither chalky nor had a glassy sheen. If successful, he would best centuries of European scientists and artisans attempting the same, and his pottery would rival Chinese porcelain as the greatest innovation in ceramic history. But that day, his experiments failed. Keeping a close eye on the oval medallions he fired and the vases he threw, he was disappointed again as they cracked and collapsed in the heat of the kiln. "We must have our Hobby Horses," Wedgwood thought, "and mount him again if he throws us ten times a day."[1]

His attitude of determination suggested a brute-force approach, a blind, even random search, which has led historians of technology to debate whether Wedgwood was a scientist. One sees in his methods "sufficient evidence to justify regarding Wedgwood as a true scientist of his age."[2] Another denies that label because Wedgwood's "mental processes and his industrial processes were based on a deeply empirical approach to problem solving," noting that his work contains "no successful application of scientific theory to a technical problem."[3] A third splits the difference by labeling Wedgwood's thought as "protoscientific thinking."[4] While it is fascinating that these "gatekeepers" protect the idea of what it means to be a scientist, they all miss the point. Wedgwood was an engineer, what we would call today a materials engineer, and the proof is in his mastery of the three key strategies of the engineering method: applying trial and error, building on past knowledge, and accepting trade-offs.

Josiah Wedgwood, born in 1730, was the youngest of thirteen children, part of the fourth generation of Wedgwood potters. His father owned the Churchyard House and Pot Works, where he earned a modest living, just enough to buy a family pew for £7 (about $1800 today) at the parish church. When his father died in 1739, Josiah's oldest brother inherited the business, and at age fourteen, Josiah became an apprentice to his brother to learn the "Art of Throwing and Handling."[5] He mastered the art despite being unable to operate the kick that turned the thrower's wheel thanks to a childhood bout of smallpox, which left him with a pained and weakened right knee.

As Wedgwood grew up, so did the industry. When he was born, a typical pottery in Staffordshire was owned by a man with a single shed, a vat for mixing clay with water, a sun pan for evaporation, a single oven for baking, and one or two workers. Often mother and daughter loaded the pottery into baskets carried by mule to nearby markets, or the potter sold to itinerant salesmen who loaded them in panniers on the backs of donkeys and hawked the wares in market towns. Rarely did they reach London; never did they sell outside Great Britain. But the demand for pottery increased throughout the seventeenth century, driven by the developing taste for tea, coffee, and chocolate—hot drinks best sipped from inert ceramics rather than metallic-tasting pewter. The wealthy used, as replacements for pewter dishes, expensive porcelain from the East, Continental faience, or delftware from the Netherlands. But for the middle class, the eighteenth-century British pottery industry developed creamware.

Creamware is a clay body ceramic, fired at a low temperature, between approximately 1110°F and 2010°F—a much lower temperature than that used to create the masterpieces of the Far East—with a high porosity that causes it to thud when struck, in contrast to the bright ring of porcelain. For color and texture, the potters of the Staffordshire region settled on light clays from the West Country to create a biscuit- or buff-colored creamware coated with a clear glaze of lead ore. Wedgwood captured the appeal of this ceramic when he described it as "a species of earthenware for the table quite new in its appearance, covered with a rich and brilliant glaze, bearing sudden alternations of heat and cold, manufactured with ease and expedition, and consequently cheap having every requisite for the purpose intended."[6]

Wedgwood built on this insight when in 1759, at age twenty-nine, he set up his own pottery works. "I saw," he later wrote with a gardener's analogy, "that the field was spacious, and the soil good, as to promise an ample recompense to anyone who should labor diligently in its cultivation."[7] He rented a small pottery for £15 (roughly $3500 today), hired his cousin for £22 (or about $5000 in modern currency) a year, and set out to refine creamware—"to try of some solid improvement as well in the *Body*, as the *Glazes*, and the *Colors*, and the *Form*" of the creamware dishes.[8]

For over a decade, Wedgwood executed a set of careful experiments to perfect creamware, to remove any variability in its manufacture, especially its unpredictable color after firing: his early creamware varied from a light primrose to straw to a deep saffron. When working on a new product, he said of himself, "the foxhunter does not enjoy more pleasure from the chase, than I do from the prosecution of my experiments."[9] In a large notebook, he recorded every experiment. The numbered entries described every result, all noted with care in his bold, clear cursive script: "This seemed to separate; part is run thin like water and is a good color, or rather no color at all; the other thicker part is greenish."[10] If an experiment turned out well, he wrote in large letters, often with an exclamation point at the end.

Wedgwood rocketed to fame from a single order when, in 1765, Queen Charlotte, wife of King George III, ordered a twelve-place-setting tea set. The astute Wedgwood shortly renamed his creamware "Queen's Ware," and for the rest of his life, he began his ads with "Josiah Wedgwood, Potter to Her Majesty." Sales of Wedgwood's creamware soared. Now, instead of a few itinerant peddlers, hundreds of packhorses streamed from the Staffordshire potteries, each horse carrying in its panniers up to two hundred pounds of rattling plates, cups, bowls, saucers, teapots, jugs, weights, mortar-and-pestle sets, and the Queen's Ware tiles that lined bathrooms and sewers throughout England. These "pot wagons" took two days to travel the sixty miles to Liverpool, and once at the port city, Wedgwood's creamware was exported to North America, the West Indian colonies, and every port in Europe.

Although his first success was these "useful wares," as Wedgwood called them, he saw another market for his ceramics: ornaments, such as cameos, medallions, and decorative vases. For this, the buff-colored creamware would not work; he needed a white ceramic. Nor would the rough earthenware finish of the creamware be appropriate; he needed a smooth stoneware.

Like any potter, Wedgwood turned his thoughts to porcelain, the holy grail of the industry. Porcelain was, since at least 1000 CE, prized for its snowy, translucent, ultrasmooth surface, a surface that feels cool when touched and that rings with a bell-like tone when tapped with a tuning fork. Wedgwood knew the secrets of porcelain manufacture; he had a copy of de Milly's *Treatise on Porcelain*, which detailed all methods—from the "true" porcelains using China's soft, white kaolin clay to the artificial, or "soft paste," versions using white-burning clay and ground glass. He also understood the process used by the nearby Chelsea and Bow pottery works, the latter of which owned the patent on the Cornwall clay suitable for true porcelain. But Wedgwood scorned the method used by both Chelsea and Bow as "one of the worst processes."[11] For his own ends, he dispatched a scout in 1767 to bring him five tons of "Cherokee earth"—a clay that behaved in the kiln like Chinese kaolin—from North America at the enormous cost of £615 (roughly $120,000 today). Wedgwood's plan to create a rival porcelain was to build on the time-honored way of making porcelain, empowered by a strategy of trial and error.

When used in everyday conversation, "trial and error" is rarely meant to inspire or impress. We think of it as a kind of last resort when more effective and efficient means have been ruled out and all that's left is the tedium of repeatedly testing solutions to see what works while spending most of our time seeing what doesn't. But Wedgwood's meticulous record keeping and reapplication of established knowledge are key to the trial-and-error strategy as a powerful problem-solving tool. It's not a blind search but, as we see with Wedgwood, a systematic exploration of a design space, where an engineer varies the value of design parameters within that space and records every possible useful variable.

In these trials, the results are initially unknown, but one isn't driving at the solution through the brute force of a desperate, wild, tedious search. The method is not random. It is guided by intuition. Despite how soft the word "intuition" might seem, for the engineer, it's a necessary, tangible skill developed in part by a career of problem-solving experience and record keeping but also by a reliance on a well of past knowledge. Rules of thumb, as always, are the key institutional knowledge passed down between eras and generations, but also useful for Wedgwood was the knowledge of what wouldn't work when attempting to create porcelain and the hints at materials, processes, and techniques that might be made to work better. Past knowledge sets up guardrails that tell problem-solvers when a possible solution is going off-road and thus helps ensure that the failures they encounter will provide meaningful feedback that can be looped back into their next round of trials. By considering his forebears and contemporaries, Wedgwood was posting the guardrails on his path. In this way, a skilled engineer can be called a kind of "conservative," not in a political sense but in the broader definition of looking to preserve the functional solutions of the present and past while making cautiously incremental adjustments—just enough to solve their particular problem at hand—that make sure attempted solutions don't veer into uncharted territory where oversights can have real consequences in the real world. They know that the best results come from making small changes to the state of the art, while a radical engineer risks building a bridge that will collapse. An intuition constructed from records, experience, and institutional knowledge, like rules of thumb, never guarantees success, but it does point the engineer toward the trials and errors that are most likely to produce useful results and deepen the collective well of knowledge.

Beyond the chemistry of its manufacture, Wedgwood understood the financial problems inherent in making and selling porcelain. The Chelsea and Bow factories struggled to be solvent and, as Wedgwood predicted, went bankrupt in a few years. The dominant European "soft paste" porcelain maker in France used a process so complex that it drained the royal treasury. And the memory of a fellow potter, Nicholas Crisp, spooked Wedgwood. "Poor Crisp haunts my imagination," Wedgwood said, "continually—Ever pursing—just upon the verge of overtaking—but never in possession of his favorite object."[12] Wedgwood's experiments with porcelain never satisfied him, but before he could refine them, an accident and a chance meeting supercharged his development of ornamental wares.

In the spring of 1762, Wedgwood rode on horseback to Liverpool, although his weakened knee ached. Along one of the narrow lanes in Warrington, Wedgwood swerved to avoid the wheels of an approaching cart and crushed his leg against a wall. He continued the twenty miles into Liverpool, but the injury prevented his return home. To grant himself time to recover, he stayed at the Dale Street Inn under the care of Matthew Turner, a local surgeon whose broad interests—chemistry, classics, and the arts—attracted a group of like-minded friends. Among these men of high society was Thomas Bentley, a Liverpool merchant with whom Wedgwood began to form an easy friendship, their mutual attraction fueled by the fact that each had what the other lacked.

Bentley was from the city. He was a quiet, educated, sophisticated merchant and man-about-town. The self-made, rural-raised Wedgwood was a blunt, forceful industrialist potter, tied to, as he said of his pottery works, "this barren plot of Earth."[13] Bentley had refined features and a graceful, relaxed presence. Wedgwood, a large, jowly man, was never at rest—in a portrait by the masterly Sir Joshua Reynolds, Wedgwood seems eager to rise and stride out of the frame. Now, confined in Liverpool, the bored Wedgwood delighted in daily visits from Bentley. As they smoked pipes, they talked about science, religion, poetry, logic, politics, pottery, and ways to improve the roads and canals. To stay in contact after Wedgwood returned to his pottery works in Staffordshire, they wrote to each other often, while Wedgwood's wife urged Bentley to travel to the works on horseback. By her reckoning, his "fat sides require a good deal of shaking."[14] The refined Bentley traveled to the works, of course, by coach.

Later, in May 1768, Wedgwood would decide to have his weak right leg amputated, an event for which Bentley was the only friend in attendance. A local surgeon, probably assisted by Erasmus Darwin (grandfather of Charles), cut the leg off, while Wedgwood's only relief from the pain was doses of laudanum. For the rest of his life, he wore a wooden leg; he kept a closetful to replace the many legs he wore out.

Not long after their meeting, Wedgwood invited Bentley to become his business partner. Bentley resisted. His expertise, he said, was in the woolen and cotton trades, not pottery. But Wedgwood would not be deterred. In a long letter of November 1766, he spelled out an expansive vision of what they could create together. They would start, he wrote, with "root flower pots of various sorts ornamented & plain. Essence pots, Bough pots, flowerpots, and Cornucopias. Vases and ornaments of various sizes, colours, mixtures & forms, ad infinitum," then design statues of "Fish, Fowl, and Beasts, with two legged Animals in various attitudes" and "Ten thousand other substantial forms, that neither you nor I, nor anybody else, know anything of at present."[15] By May 1766, Bentley relented, and the two men formed a partnership.

For their business, Bentley developed the commercial contacts necessary for selling their new wares and gained knowledge of the market. Wedgwood drove technological innovation and managed their factory in the north of England—"content with fashioning my clay," said Wedgwood, "at a humble distance."[16] To sell their products, Bentley moved from Liverpool to London, where, as a clubman, he consorted with royalty and their ladies-in-waiting. Through his connections, he became friends with Joseph Priestley, a chemist who discovered oxygen; James Watt and his partner, Matthew Boulton, who pioneered steam engines; Nevil Maskelyne, the royal astronomer; painter Sir Joshua Reynolds; Sir William Hamilton, a prolific collector of antiquities; and Benjamin Franklin.

Bentley's letters from London arrived at Wedgwood's home and pottery works in the north of England three times a week—two thousand letters in thirteen years. He filled them with the latest world news before it hit the local newspapers: the king of Naples was robbed of his watch and diamond rings worth four thousand ducats "by some villains who broke into his closet," and the ship *Susannah* of Amsterdam, laden with sugar, coffee, cotton, and cocoa, crashed on the rocks near Barnstaple in Devonshire.[17] But most importantly, they contained insight into the London fashion scene.

In the late 1760s, a new craze was sparked by the discovery of Pompeii and Herculaneum. Although they had been excavated thirty years or so earlier, the publication of illustrated books detailing the finds fueled the imagination of the British public. One popular title featured sumptuous color engravings of the antiquities that Sir William Hamilton, as the British ambassador to Naples, had collected for more than thirty-five years.[18] When the leisured class of the eighteenth century flipped through these pages, they saw in the vases and plaques the birth of European civilization that, to them, was a golden age of stability—an age that contrasted with the rapid changes of industrializing Britain and the growing tension with the American colonies.

Wedgwood borrowed a copy of Hamilton's book, probably from Bentley, to inspire his next line of ceramics. The result was figurines, plaques, cameos, seals, portrait medallions, and ornamental candlesticks created in either a black ceramic that could be polished to look like bronze or cast in terra-cotta with an agate or variegated finish to suggest natural stone. Wedgwood decorated both types of ceramics in the "encaustic" style to imitate Greek vessels: vivid, rich paintings of scenes from Greek history and myth. This style of painting was so important to their business that Wedgwood patented it in 1769. Yet only a few years later, this style was on the decline.

As Bentley consorted with the aristocracy of London, he sensed another trend: a brightening of the world. He noticed in the work of architect and interior designer Robert Adam, who earned large sums to redesign the rooms of his wealthy clients, a bold use of color. A typical eighteenth-century ceiling was white, but Adam surprised with the rich crimson ceiling at Northumberland House. And at Syon House in West London, Adam decorated with a range of colors: white and green chimney pieces; gilded decorations set against blue and green backgrounds; and blue, green, red, and gold floors. Yet the results were never gaudy—color was always tempered with delicacy in Adam's decoration. Compared with these bright interiors, Wedgwood's black medallions, cameos, and vases, the latter with garish drawings, seemed out of step with Adam's subtle yet colorful designs. This prompted Bentley to ask Wedgwood, in late December 1772, for a white "finer body for gems"—a smooth, hard ceramic for their prized cameos and medallions.[19]

In Wedgwood's response a few days later, he assumed that Bentley was asking him for porcelain, whose production he had not completely mastered. "I think a China body," as Wedgwood called it, "would not do. I have several times mixed bodys for this purpose, but some have miscarried, & others have been lost or spild for want of my being able to attend to, and go thro' with experiments."[20] He hesitated, though, to develop a new ceramic. "At present I cannot promise to ingage in a course of experiments—I feel that close application will not do for me—If I am stronger in the spring, something may be done, but at present my health is in too delicate a situation & my life I believe may, at least be set down *Double hazardous.*"[21] Over the summer, Wedgwood had lost so much weight that his clothes only fit when altered by several inches. His vision also often bothered him. "My eyes," he complained to Bentley, "have been in such a way for two or three days past that I durst not look upon paper. They had quivering objects before them."[22] These were likely what we call today "vitreous floaters," which are irritating but benign. Wedgwood's doctor, Erasmus Darwin, declined to prescribe any "Physick." Instead he recommended to his perhaps hypochondriac patient, "to live pretty well, to take moderate exercise & to keep free from care & anxiety."[23]

Wedgwood could "live pretty well" in his thirty-four-room mansion, Etruria Hall.[24] The library was stacked floor to ceiling with books, so many that Wedgwood's wife insisted he buy no more until he built another house. He could continue his study of the chemist Joseph Priestley's work, the theories of humankind and society proposed by Jean-Jacques Rousseau, or the moral philosophy of Richard Price, an opponent of slavery and supporter of American independence, as was Wedgwood himself. The most famous of Wedgwood's medallions featured a kneeling slave in chains with the appeal "AM I NOT A MAN AND A BROTHER?" embossed around the edge of the ceramic gem.

Or he could relax in the mansion's large salon. He might organize his fossil collection, an enthusiasm since childhood (indulged as an adult by blowing up with gunpowder the nearby sandstone to uncover fossils). He enjoyed hearing Darwin describe his fanciful design for a windmill to grind colors and relished their conversations on "electric attractions" after both read a book on the subject. At times, Wedgwood met with Joseph Priestley to discuss their common interest in experimentation, although Wedgwood might well inquire about what Priestley meant by a "battery," a device that puzzled Wedgwood.[25]

A few weeks later, Wedgwood reported that he no longer lost weight and that his mood improved—he could work again "by casting care behind me as much as may be."[26] Wedgwood returned to his workbench, eager to create that white "finer body" ceramic. He reported to Bentley "some promising experimts lately upon fine bodies for Gems & other things."[27] Yet after February 1773, Wedgwood faced another interruption. Bentley's success as a salesman had taken away Wedgwood's time with an order from, in Wedgwood's words, "my Great Patroness in the North."[28] The cosmopolitan and well-connected Bentley procured, through a friendship with the Russian consul in London, an order from Catherine the Great of Russia for 944 pieces of custom creamware—680 pieces for dinner service and a dessert service of 264 pieces. Each dish was decorated, at Catherine's request, with a British landscape, antiquity, or garden. "I love English gardens," she wrote to Voltaire, "to the point of folly."[29]

Wedgwood's wares for the Russian table service were buff colored, and the scenes of landscapes, antiquities, and gardens were printed in a black ink, tinged with purple, using a great innovation of English ceramics. With a flexible sheet of gelatin, workers transferred images from an engraved copper plate onto a piece of pottery. Surrounding each scene at the center of the plate was a border—oak leaves and acorns for the dinner service and ivy for the dessert— broken at the top by an escutcheon encircling a splayed, bright-green frog! Although Wedgwood did not like the frog emblem, Catherine insisted on its placement on every dish. She intended this service for Chesme Palace, at one time called La Grenouillère or Kekerekeksinen because it sat on a frog marsh. This emblem earned the set its nickname the "Frog Service," a title still used.

The Frog Service was manufactured at a new factory, built in 1767, but only now fully coming on line. Wedgwood christened his new factory "Etruria Works," under the misconception that the ancient Greek and Italian pottery discovered by Hamilton and others was Etruscan. This seven-acre complex of buildings was bordered by walls on three sides and, by 1777, a canal on the other. When the bell in the cupola of the central factory building rang at 5:45 a.m., three hundred workers dressed in white aprons flowed into the works from the adjoining village, built to house the factory workforce— tidy homes, each with iron steps, six inches wide, always brightly polished. Soon the factory's windmill and waterwheel started turning, ready to lift and turn the heavy chert stone held by stout oak beams. The stone crushed and then ground one of the many piles of raw clay that had been weathered for nearly a year—china clay from Cornwall, ball clay from Dorset, flint stone from France. The factory soon filled with the acrid smell of burned coal from the bottle ovens. As the day progressed, the clay traveled in an efficient semi-circle from the pile outside the works, through the works, to the ship house and the packing house.

Keeping the works going and the workers in line occupied much of Wedgwood's time. He fought a constant battle against stoppages for a wake or a fair and against drunkenness, including the periodic three-day drinking binge—an accepted behavior at other potteries. To their employer, the Etruria workers were little better than filthy, unruly, and wasteful incompetents, and he doubted their ability to direct themselves. To remedy this, he dominated the works. His mansion, Etruria Hall, stood on a ridge just above the factory, only a quarter mile away. Each day, Wedgwood visited the factory dressed in his short, curled bob wig, lace frill, blue collarless coat, scarlet vest laced with gold and with gilt buttons, a cocked hat, and, if it was a dress celebration, a sword. At the factory, he imposed a rigid set of rules: a ban on drinking, standards of care and cleanliness, and rules to avoid wastage. To prevent broken windows, Wedgwood ordered that "any person playing at fives"—a kind of handball—"against any of the walls where there are windows forfeits 2 shillings," a substantial sum for a worker.[30] When workers arrived at the main gate, the "clerk of the manufactory" recorded their arrival with color-coded chalk on a blackboard. Anyone caught scaling the walls to avoid being marked tardy was fined.

With the Frog Service, Wedgwood and Bentley decided to beef up their marketing by opening a new showroom in London to show off the collection before it was shipped to Russia. To draw the attention of the urban aristocracy, they set up their new showroom on Greek Street, in the fashionable and elite Soho district, where nearby venues attracted London's most reliable sources of disposable wealth. Less than half a mile away was the newly built Pantheon, a building as Roman as could be created in the eighteenth century. Under its giant dome—copied from that of Rome's Pantheon, although in plaster on a flimsy wood framing—occurred subscription-only concerts (red tickets for ladies and black tickets for gentlemen) and elegant masked balls. The most extravagant masquerades, though, were those at Carlisle House in Soho. There, a Mrs. Cornelys—a lover of Casanova, also the father of her daughter—attracted royalty like the Duke of Gloucester and the king of Denmark.

When the showroom opened on June 1, 1774, it smashed attendance records. Queen Charlotte attended, as did her brother, His Royal Highness Ernest of Mecklenburg, and the king and queen of Sweden. For months, the fashionable of London thronged the rooms and blocked the street with their carriages. If any of those visiting wanted the wares, Wedgwood and Bentley were happy to sell reproductions, without the frog, of course.

In mid-1774, with Catherine the Great's order done and the showroom a success, Wedgwood retreated to his study to work on that "finer body for Gems" requested by Bentley two years earlier. Having rejected porcelain, he reviewed the "promising experimts" he had mentioned to Bentley in February 1773. Wedgwood, the keen gardener, called these early tries his "roots" and "seeds," which he hoped "will open & branch out wonderfully."[31] Once finished with his notebook, he opened a trapdoor in his study and descended, careful of his wooden leg, a narrow set of brick stairs to the cellar.

He entered a large room, one of many rooms separated by large, arched wooden doors on heavy iron hinges. In a corner was a table for his experiments, a well for drawing in water, and bins and troughs for his vast variety of raw materials. He had stored nearly one hundred substances, which composed his palette for creating ceramics, and if he needed anything else in particular, he could use the secret tunnel—six feet in diameter—that connected Etruria Hall to the works. Near the steps he had descended, there was a door that opened to the outside, but it was hidden by a double-walled screen that formed a winding passage. This allowed barrels and boxes to be brought in but prevented anyone from seeing inside, a reaction to Wedgwood's nearly irrational fear of industrial spies. He worried about his "antagonists,"[32] as he called his competitors, and feared their "arcanists"[33]—a term for workers with knowledge of the trade secrets of ceramic manufacture, especially porcelain, but eventually used to describe those who bartered their knowledge of ceramic secrets for employment, then stole the secrets of their new employer.

Wedgwood chose as a base white clay from the Isle of Purbeck, created thanks to feldspar, a mineral rich in silicon, once lodged in the granite of Devon and Cornwall, whose rivers had eroded and washed downstream millions of years before. Next, he selected a blindingly white powder called spath fusible, or barium carbonate. Wedgwood carefully mixed the finely ground spath with the white clay and a small amount of calcined flint. The flint, a common ingredient for ceramics, increased the whiteness of the body and let him fire it above 2,200°F, a temperature that vitrified some of the spath, turning it into a glass-like substance. Wedgwood shaped the mixture into thin slabs that he could hold in his palm and, with metal type, stamped into each a three- or four-digit number to identify them later. On some, he impressed the letters "TTBO," which meant "tip-top of biscuit oven," the hottest part of the oven. On many, he scratched into the clay the ratio of spath to clay: 35 to 1, 30 to 1, or 20 to 1. "Preparing the Clay," he once explained, "is the first business in order of time, & ought to have very particular attention paid to it as the beauty & value of the manufacture must in great measure depend upon the mixture and preparation of the body."[34]

Throughout the next four months, Wedgwood fired hundreds of trials. But the results disappointed him. "At one time," he reported, "the body is white and fine as it should be," but the next trial resulted in a cinnamon color.[35] Some samples melted to a glassy finish, while others were as "dry as a Tobacco pipe."[36] He wrote in despair to Bentley, "I am almost crazy," wishing he had "more *time*, more *hands*, & more *heads*."[37] And he hoped for a reprieve from visitors: "A man who is in the midst of a course of experiments should not be home to anything or anybody."[38] As a famous potter and entrepreneur, he was bothered by all manner of people. A "Mr. Earl" of Yorkshire stopped by with a cameo, claiming it to be part of the collection of the late Cardinal Albani, a leading collector of antiquities and art patron in Rome. Most of the samples Wedgwood found "very poor indeed," because the pieces were cracked and had missing sections.[39] A Dr. Percival of Manchester dropped off "eight very close printed Quarto pages" on the *Miscellaneous observations & experiments on the poison of Lead.*"[40]

By August 1774, Wedgwood reported to Bentley, "I have now begun a series of experiments upon materials which are easy to be had in sufficient quantities, & of qualities allways the same."[41] This substance, called cauk, was new to him. He had only recently come across this material on a visit to mines near Derbyshire. He now added cauk—the blindingly white mineral barium sulfate—to his mixture. This white chalky substance, he thought, might serve as a partial replacement for the spath fusible. Cauk in its raw form also contained small lumps of lead, which Wedgwood carefully removed. We see here why trial and error is not just a random search. The steps are guided by creative and evolving theoretical ideas—in this case a hypothesis about the importance of cauk—about what properties to measure, the possible impact of the parameters, and which parameters to vary.

From a shelf, he selected a mold of Mrs. Matilda Fielding, the daughter of the royal governess to the children of King George III and Queen Charlotte, with which to make a cameo. After molding the white relief portrait, he mounted it on an oval of the same material enameled in purple, then baked them so the portrait and oval were fastened. The cameo, a mere inch and three-quarters long, constituted a partial victory. But he still had inconsistent results. "I cannot work miracles in altering the properties of these subtle and complicated materials."[42]

Encouraged, though, by this partial success, Wedgwood tried to fire larger bas-reliefs, but the surface of the medallions blistered into unsightly specimens in the kiln. Suspecting the spath fusible, he decided to replace it entirely with cauk. The result was a heavy, strong, impervious white stoneware that fired to translucence with a uniform waxen surface—the blistering, shrinking, and inconsistency were gone. In the crucible of the kiln, half the cauk reacted with the clay to form a network of barium, aluminum, silicon, and oxygen, while the other half baked into glassy barium sulfate particles. With victory so near, Wedgwood then added a last note of perfection to the mixture. A substance that, when combined with his new ceramic body, revolutionized the world of ceramics: smalt.

Smalt was often used by early seventeenth-century painters to provide an inexpensive substitute for ultramarine, or, as in Rembrandt's later paintings, to increase the volume and texture to paints, to deepen their colors, and to cause the paint to dry faster.[43] Its key ingredient was cobalt oxide, mined in Saxony—both the sole source of the ore that contained the chemical compound and the region in which smuggling the precious ore was punishable by death. Wedgwood got his from a man who kept a warehouse near Hamburg. "He had found out the way of procuring it," Wedgwood reported, "by means of some Jews, one of whom they say has been discover'd hang'd for the practice."[44] The ore was roasted, then melted with quartz and potassium to create a deep blue-black glass, which was ground to a fine powder pigment. Although it was often used by potters to apply color to the exterior of a piece, Wedgwood instead intended to add smalt to his revolutionary clay mixture. Infused within the stony body of his burgeoning ceramic, this addition produced a regal note of pale blue.

In December 1774, Wedgwood wrote with enthusiasm to Bentley about his discoveries, claiming that with this new ceramic, "you will make the finest things in Europe," and it will be "the ultimate perfection of our Cameos."[45] He was confident he could create any size from "the smallest gem for rings" to the "Herculaneum size" plaques, ovals fifteen inches by twelve inches.[46] He was so certain of the composition of this new ceramic that by early January 1775, he asked for Bentley's help in hiding the source of the raw materials.

He feared that "our Antagonists" might overtake them, so they needed to "leave them behind."[47] He perhaps feared for good reason, as all the raw materials for his invention were from sources near his Staffordshire pottery works, most from Derbyshire. To throw potential spies off the trail, he suggested that Bentley order shipments of his raw materials to London, then "disguise it there & send it to me hither in boxes and Casks."[48] Four days later, he changed these plans. Obsessed with "procuring & conveying *in Cog*"—meaning *incognito*—"*the raw material*," he told Bentley to ship it to the West Indies, then have it shipped back to London.[49] "We shall leave them at so great a distance," wrote Wedgwood, "& they have so many obstacles to surmount before they can come up to us, that I think we have little to apprehend on that account."[50]

Yet despite his confidence, his tests failed. For nearly three years, from May 1775 to November 1777, Wedgwood struggled. In the kiln, his medallions and cameos cracked and his vases collapsed. He noticed, though, a curious discrepancy. While a finely ground mixture of his new clay—one that yielded a smooth, waxen, unglazed surface—would collapse in the kiln, a vase composed of coarse particles stood up to firing. His breakthrough occurred when he employed a technique used by no other potter of the time. He created a sturdy vase from the coarse material, fired it, and then dipped the rough body of the cooled piece in a solution of the finely ground mixture of the same smalt color to coat it with a thin ceramic shell. When he fired the vase again, the outer coat solidified into a smooth, stony sheen, and the vase held its shape. The inner body had the solid heft of barium sulfate, a delightful ping when struck, and the outside had the pleasant texture of fine feldspar silica. Wedgwood had created a new ceramic to adorn the homes and accessories of royalty and aristocracy around the world: jasperware or jasper for short. Even the most expert of collectors did not detect Wedgwood's sleight of hand, which remained a trade secret for over two hundred years—not even his family members knew this secret.

His breakthrough, this "dipping" to create a veneer, illustrates the last of the three strategies used by an engineer: trade-offs. Any engineered design has limitations, so when designing a product, an engineer must decide how much to balance the particular characteristics of an object, a balance closely connected with the notion of "best." In implementing a trade-off, an engineer must balance between two incommensurable aspects of an object—you cannot have *both* completely. A classic example is the cylindrical aluminum soda can: consumers and retailers would prefer cuboid cans because they can be easily stacked and stored with no wasted space, but the sharp corners of the can are weak; the strongest can, which uses the least amount of material for a given volume, is a sphere—no sharp edge, only curved surfaces. The spherical can, of course, is a consumer nightmare: it rolls off the table. An engineer trades off a bit of each desired characteristic to find a balance: a can is cylindrical, so it stacks and stores a bit like a cuboid can yet has round sides like a sphere to avoid the weakness of corners. Wedgwood faced the same type of choice: he wanted the sturdy ceramic body, but if he stuck with widely accepted glazing techniques, he wouldn't get the ideal uncracked exterior. So to compromise, he found a method that was best given his parameters. His solution of a veneer allowed him to have a bit of both—the heft of a sturdy ceramic body with a smooth exterior—even though it would potentially put off buyers and necessitated secrecy.

With jasper now refined, after five thousand recorded experiments, Wedgwood created ceramic masterpieces. Among the most stunning is a sixteen-inch tall, light-blue vase, now on display at the Victoria and Albert Museum in London. Its surface is as smooth as that of porcelain but unglazed; the blue matte surface has a striking evenness of tint without a drop of paint or other pigment coating it.

The vase overflows with ornamentation: a white figure of Apollo, surrounded by the nine Muses, and capped with a lid featuring a rearing white Pegasus. These ornaments look soft, almost like ivory or vellum, with neither a chalky nor a glassy sheen. Their thin edges are translucent so the blue shines through, suggesting the ornaments are drapery on the vase. The folds in the Muses' robes look like billowing cloth, and Pegasus's wings seem ready to flutter.

But before marketing such masterpieces, Wedgwood tried once more to hide the ingredients of his new ceramic. He reminded Bentley to get cauk from Liverpool, have it shipped to London, and explain to the supplier that he "hopes of opening an exportation trade for it," implying that it was to be exported and not sent north to Wedgwood's factory.[51] And finally, he asked Bentley to spread the word that the secret to jasper was Cherokee clay—the kind he had ordered five tons of in 1767—thus hoping to confuse his "antagonists."

With jasper, Wedgwood achieved a global distribution. He exported 80 percent of his wares: Dublin, Dunkirk, Nice, Paris, Amsterdam, Brunswick, Dresden, Leipzig, Naples, Venice, Malaga, Moscow, St. Petersburg. By 1795, Wedgwood sold his pottery to every regal house in Europe. Bentley suggested Chinese markets, but Wedgwood worried they would copy his designs in porcelain, although he did ship seven vases to the Qianlong emperor in 1793.

Today you can still buy jasper, which has been manufactured continuously since 1780, except for a short time in the nineteenth century. None of the jasper ever had an ounce of Cherokee clay, though Wedgwood's grandson reported, about fifty years after jasper first appeared, that "two arkfuls" remained at Etruria Hall.[52]

Wedgwood's work in the cellar of Etruria Hall illustrates superbly the three key strategies of the engineering mindset—using trial and error, building on past knowledge, and embracing trade-offs. These strategies are still essential today. We can clearly see them in the work of twenty-first-century engineers. Frances Arnold, as described in chapter 3, tailored enzymes for industrial application by using trial and error, building on past knowledge in being guided by nature, and embracing a trade-off between perfection and cost— she could have continued forever to refine the enzyme's activity but settled for good enough. Both Wedgwood and Arnold, separated by 226 years, have been considered through the lens of science: Arnold with a Nobel Prize for Chemistry, and long debates about whether Wedgwood was or was not a scientist, though he has never been labeled by a historian of technology as an engineer. One look at his chemicals, furnaces, and experimentation, and they immediately screamed "scientist!" This confusion highlights a question: What is the relationship between science and engineering?

So far, we've observed that engineers often move forward before scientific understanding. This observation demolishes the notion of engineering as applied science; to view engineering as applied science conjures an image of science as an organized battlefront that expands, conquers all uncertainty, and enables technological marvels. This view is reinforced by the observation that extraordinary growth in scientific knowledge always coincides with rapid technological advance. Yet a more accurate picture is of engineers fighting a guerrilla war to change the world, combating scientific uncertainty with whatever tools and techniques work, those rules of thumb. Although scientific understanding is irrelevant to engineering's existence, scientific knowledge, as explored in the next chapter, has the power to supercharge engineering. The relationship between science and engineering is well illustrated by a nineteenth-century invention still central to our age, the greatest gift from the Victorian inheritance: the steam turbine, a kind of engine that changed the world and still lights our homes.

6

SCIENCE

The Gold Standard for Rules of Thumb

Charles Parsons left the deck of the *Turbinia* and descended to its engine room the moment he heard the signal gun. Soon, the royal yacht carrying the Prince of Wales, Queen Victoria's son, would pass by his little ship on its way to inspect the British fleet. As Parsons entered his ship's crowded engine room, six stokers slowly fed coal to the ship's boiler.[1] Parsons shed an oilskin topcoat and settled in at a panel of gauges and valves. From this station, he could adjust the boiler pressure to control the steam flow rate through a tangle of pipes that led to his revolutionary steam-powered engine. He fixed his gaze on the pointer of a marine telegraph, labeled with *full speed*, *stop*, and *reverse*, ignoring the cacophony of the engineering room—the blast of the furnace, the beat of pumps, and the hum of a fan. Above deck, the pilot hooked his elbows through the spokes of the steering wheel, and the coxswain at the bridge clutched the rope of the danger whistle, ready to signal if the ship must be stopped abruptly. The crew was ready to demonstrate Parsons's invention—and earn its due recognition—by crashing the biggest event of Queen Victoria's diamond jubilee celebration.

Parsons's engine was a turbine, a vortex of paddle wheels driven by a jet of high-pressure steam to turn a shaft and, in Parsons's application, propel a ship. Simple in concept, yet even at the height of the Victorian era, a successful turbine had never been built—none of the more than two hundred patented designs of a steam turbine worked. To most, the idea of a turbine steam engine was an amusing pipe dream. The steam engine had been so refined in the 120 years since James Watt and Mathew Boulton's innovations accelerated Britain's first Industrial Revolution that few engineers or scientists thought it could be improved or was worth improving. Others thought the internal combustion engine would relegate the steam engine to "a curiosity to be found in a museum."[2] Yet the elegance of a steam-powered turbine kept a generation of engineers believing that the technology could be rejuvenated. To harness steam power with turbine engines meant directing the energy of expanding steam with fewer moving parts and thus less maintenance, lower weight of machinery per pound of horsepower, less coal consumption than a conventional steam engine, and nearly no vibration—a constant source of discomfort to passengers of steamboats caused by the oscillating pistons of the reciprocating steam engines used on the commercial liners of the day. Now with a new working turbine, Parsons was ready to claim his engine superior to the engines currently used on ocean liners.

In April 1897, Parsons presented details of his engine at the Royal Institution of Naval Architects. The institution's membership was restricted to professional ship architects and builders and included naval admirals, engineers from Belfast shipbuilding firm Harland and Wolff, shipbuilders from White Star Line and

Cunard—all in all, a group of professionals whom Parsons needed to impress. Unlike previous attempts, his particular version of the turbine, Parsons explained, succeeded "chiefly on the basis of the data of physicists."[3] He described the engine's exceptional performance on the *Turbinia*, specially built by him to test the engine. But Parsons's claims generated only unimpressed shrugs.[4] John Corry, who operated the Star Line, a fleet of ships that transported cargo around the world, called Parsons's presentation "too meagre with regard to construction and details," adding that the claims presented were "too premature." He doubted the lessons learned in fitting the turbine to a small ship like the *Turbinia* would apply to the kinds of large ships that he and his colleagues were interested in outfitting. He concluded that the results presented by Parsons "are not sufficient to warrant this large claim of advantages." And a naval architect suggested that Parsons remove the *Turbinia*'s turbine, slap in a reciprocating engine, and let a conventional shipbuilder "see if they can beat the results." The consensus was summarized by the *Times*: Parsons's turbine was "in a purely experimental, perhaps almost embryo, stage."[5] This dashed Parsons's hopes for adoption by the British Navy and the large shipbuilding firms, who, with such poor reviews, would never risk fortunes in the millions to design a ship around his new engine. They would remain happy with the reciprocating steam engine, which for decades had powered the British fleet and mercantile ships with good service and high efficiency.

"The new invention," Parsons once said, "like a young sapling in a dense forest, struggles to grow up to maturity, but the dense shade of the older and higher trees robs it of the necessary light."[6] If only, he continued, the invention would "grow as tall as the rest all would be easy, it would then get its fair share of light and sunshine." Parsons was determined to shine the light on his invention. So this most modest, even shy, of men—he shunned all forms of publicity so much that he refused to allow the name of his company to be painted on its gates—planned to force the world to see his steam turbine in action.

To understand the action of steam power, think of a teakettle. A cup of water heated in a kettle boils and evaporates into steam, and as the steam fills the kettle, the principal ability of boiling water to generate massive amounts of power is revealed: the volume of water as steam is orders of magnitude greater than that of liquid water— one cup of water evaporates into sixteen hundred cups of steam. The boiling water causes steam pressure to build up within the kettle until the steam is pushed out a small hole in the lid with enough force to make the classic whistle that lets the user know the kettle is ready for brewing tea. In the case of the teakettle, the steam is a by-product that's cleverly turned into a convenient signal, but to extract energy from steam to drive a locomotive wheel or turn a ship's propeller, the industrializing world relied on the reciprocating engine so trusted and beloved by Parsons's skeptics.

The reciprocating steam engine has two main components. The first is a hollow metal cylinder in which a round piston is pushed back and forth by expanding steam. A boiler, akin to the teakettle, creates a rush of steam that pushes the piston to one end of the cylinder; the compressed steam, which partly condenses back into liquid water, is drained and replaced by another rush of steam that pushes the cylinder back to its original position at the other end. But this back and forth movement, called reciprocating motion, by itself is mostly useless. It was the job of the second main component—a set of pivoted arms—to convert the reciprocating motion into circular motion that can turn a shaft or a wheel. The most classic image of this arrangement is the chugging wheels of a steam locomotive that turn the reciprocating action of the engine piston into a train moving along railroad tracks.

1) High-pressure steam enters through a small valve.

2) As the steam expands and its pressure drops, it moves the piston, which causes the wheel to rotate counterclockwise.

3) When the piston reaches the other side of the cylinder, steam is injected on the other side of the piston to force the piston to the right.

4) As the left side of the cylinder fills with steam and the piston moves right, the exhausted steam on the right side is bled from the cylinder via a small valve.

5) As the piston approaches the right side of the cylinder, the sequence returns to step 1, where steam is again forced into the right side of the piston.

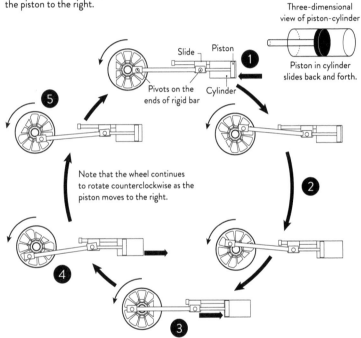

The steps of a reciprocating steam engine.

Although the reciprocating steam engine served the world well (it was the best for its time), it wastes much of the steam's energy. This is partly because once the steam has expanded to about sixteen times its original volume, it lacks the "oomph," the force, to overcome the friction of the heavy piston within the walls of the cylinder and partly because the set of arms that convert reciprocating motion into circular motion add more moving parts that use even more of the steam's energy toward overcoming friction in the engine's design rather than doing useful work. This type of engine could ultimately extract only about 70 percent of the energy contained in steam. "We take the cream off the bowl," said a nineteenth-century engineer, "and throw away the milk."[7] But Parsons wanted the whole bowl.

To create his successful turbine, Parsons combined two fields of knowledge, neither sufficient yet both necessary: first, a comprehensive cataloging of the properties of steam and a theoretical understanding of its behavior—an outcome of the bloom of scientific knowledge in the nineteenth century—and second, an engineering mind well versed in steam-powered motion.

Charles Parsons's deep, practical knowledge of steam power was learned in a childhood spent amid machines.[8] Born in 1854, Parsons grew up with imperial privilege as the son of Anglo-Irish nobility at Birr Castle, which, unlike any other castle in the British Isles, sported a foundry whose yellow flames lit the grounds at night with an eerie glow as it smelted iron and doused freshly cast ingots in the castle moat's water. Surrounding this foundry were workshops filled with lathes, cranes, and glassblowing tools, operated by a team of live-in blacksmiths. While similar manufacturing suites in the industrialized British Isles were used to feed raw materials to a world roiling with technological, economic, and political revolution, all the apparatus and human power at Birr Castle was to construct, for Parsons's father, the world's largest telescope: a seventy-two-inch reflector, known locally as the "Leviathan." The grounds always buzzed with activity as workers fashioned scientific instruments, block and tackle for the telescope, pumps, swing bridges for the moat, gates, and agricultural implements to farm the grounds. This innovative environment formed Parsons as an inventor.

Parsons recalled a childhood "making contrivances with

strings, pins, wires, wood, sealing wax, and rubber bands as motive power, making little cars, toy boats, and a submarine."[9] But none were more attractive to him than steam-powered motion. Under his father's guidance, Parsons and his brother built, in 1869, a steam carriage that traveled seven miles per hour—a stunning device in an age when the horse remained supreme for several more decades. While he was a student at Cambridge, his central enthusiasm, constructed from paper, sealing wax, and steel wire, was a prototype of a novel reciprocating steam engine called an "epicycloidal engine," defined by its four pistons arranged in a circle to reduce vibration and rotate a shaft faster than a standard-issue steam engine. It was no mere toy; he sold forty final versions of the engine to drive centrifugal pumps, and he joined a small engineering firm as a partner (a privilege for which he paid £14,000, perhaps £1 million to £2 million today) and used his engine to generate electricity for machines and electrical lighting onboard ships. Even in his home and with his family, he was always thinking of novel ways to use steam power. For his children, he designed a "spider," a small car with three wheels and a motor powered by burning rubbing alcohol that chased his children and the family dog around the lawn—his wife banned him from running such toys in the house after a miniature locomotive spit out flaming alcohol, leaving a trail of fire on the library carpet. She also forbade him from transporting the children in a steam-powered stroller of his design because she feared the cookie tin used as a boiler might explode.

At this small engineering firm, Parsons had achieved his first great commercial success as an inventor by addressing one of the main disadvantages of the reciprocal steam engine: the characteristic vibration of the engine caused fluctuations in an electrical system's voltage, which shortened the operational life of newly invented and expensive light bulbs. His impressed partners encouraged him to incrementally improve on his designs and work out the remaining efficiency problems that trains, ships, and generators had barely tolerated in steam engines for decades. But Parsons's inherited habit of toying with devices was not simply the idling pastime of a wealthy tinker. He wanted to outdo James Watt's original design with a turbine that completely rethought how we harness steam power. Like those who had theorized designs for steam turbines before him, Parsons knew there were two main methods for converting the pressurized power of steam directly into circular motion rather than through an intermediary step of reciprocal motion: one could either use the force of a jet of steam blasted like the exhaust of a rocket or rush steam through a set of paddle wheels like a steam-powered windmill.

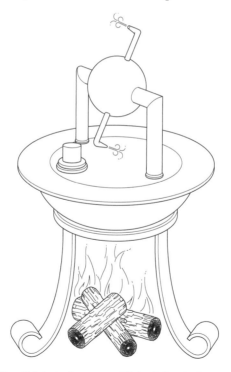

The aeolipile of Hero. Built in the first century CE, it relied on the first known use of steam to directly create rotary motion.

a.

b.

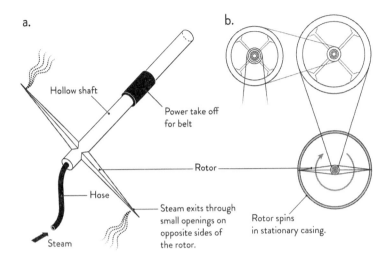

Hollow shaft

Power take off
for belt

Rotor

Hose

Steam exits through
small openings on
opposite sides of
the rotor.

Rotor spins
in stationary casing.

Steam

In the early nineteenth century, William Avery designed a steam-powered buzz saw. The saw, not shown in this drawing, was affixed to a hollow shaft. At the end of this shaft, he attached a rotor. As steam was fed through the hollow shaft, it blasted out the small holes in the rotor, which, like the aeolipile, spun the shaft.

The first was demonstrated around the first century CE when Hero of Alexandria built a small machine he called an "aeolipile," which resembled a spinning, steam-powered teakettle. This hollow sphere was filled with water and mounted on a pivot so it could rotate, and three pipes extended from the sphere. The water in the sphere was heated by a flame beneath it, and when the water turned to steam, it rushed from the pipes and spun the sphere. Using low-pressure steam that only needed to overcome atmospheric pressure to rush out of and propel the aeolipile, the device didn't have enough power to do much useful. It could, perhaps, turn a roasting spit. To generate enough torque to drive something like an engine shaft, the steam pressure must be many times greater, created by capping the pipe, then releasing it as a high-power jet when the necessary pressure is reached. But to be at all useful to the industrial barons of the British Empire, the aeolipile would need to release steam at a pressure seven times greater than that of the atmosphere, which would blast steam at 1,200 miles per hour, many times faster than the most powerful hurricane winds, and tear the aeolipile apart—a problem illustrated by a powered saw designed by William Avery in the early nineteenth century.[10] To create a buzz saw for a sawmill in New York City, he fit a circular blade on the end of a long, hollow shaft, and to turn this shaft with steam power, he fastened the other end to a five-foot-long pipe with small holes in its end caps. As he fed steam through the hollow shaft, it flowed out the small holes in the perpendicular pipe and, like the aeolipile, spun the shaft. Its rotation was so fast that it burned the lubricating oil in the bearing that held the shaft, and the saw blade blew apart, shooting shrapnel through three floors like pellets from a shotgun blast. Luckily, no one was hurt.

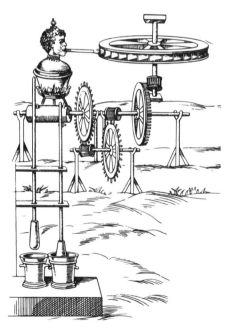

A steam-drive device envisioned by Giovanni Branca in 1629. Steam generated in a kettle was blasted from the mouth of the human figure. This steam turned a paddle wheel, which in turn drove gearing to operate pestles.

The second method historically didn't fare better. The earliest recorded version of the steam paddle wheel device is from 1629, when Giovanni Branca, an Italian engineer, proposed a giant boiler, shaped like a human head, that blasted a jet of steam onto a paddle wheel. When engineers built devices like Branca's, not only did the paddle wheel spin so fast that it blew apart, the expelled steam cut through the metal blades. It seemed that inherent in the direct transfer of steam power to circular motion was an impasse between the force of steam necessary to do useful work and the ability of a machine's materials to withstand the relentless power of that force.

To overcome this no-win scenario, then, Parsons knew he must slow the flow rate of steam to tame the rotations and, as he put it, "avoid the well-known cutting action on metal of steam at high velocity."[11] To this problem, Parsons applied what his wife called his "outstanding feature," his power of concentration.[12] "Nothing disturbed him when he was absorbed in a problem," she said. "No noises, no discomfort, no time and no meals—he was perfectly oblivious to them all." Indeed, Parsons, when on the scent of a solution, worked twenty-four, even thirty-six hours straight. Late at night, he sat in his study, a white-sulfur cockatoo at his side, to whom he absentmindedly fed sugar lumps. And during the day, when in this state, he walked around his machine shop with a glassy gaze. Once, he was so distracted that he stepped into a bucket of lubricating oil; another time, he stepped backward off a three-foot-high platform, struck the edge of a machine, and nearly broke his ribs. This deep thought, though, led him to his breakthrough design: Parsons realized that the high velocities that tore apart early machines occurred because the high-pressure steam was vented to the atmosphere, which causes the steam to rush out at hurricane speeds. He reasoned that if he could drop the pressure through a series of contained steps—so no drop was too large—he could reduce the flow rate of the steam yet eventually extract all its energy.

Although his idea was straightforward in principle, Parsons described the practical problems of executing his design as of "almost infinite complexity"[13]—the number of steps he should use to slowly drop the steam pressure, the speeds of rotation caused by the steam's pressure, and the question of whether the steam's velocity was low enough to avoid cutting steel. To find his way in this "infinite complexity," Parsons turned to what he had referred to as the "data of the physicists."[14] His engineering solution would be impossible without the help of science.

In the thirty years before Parsons's work, the field of thermodynamics—the study of energy and its transformations—blossomed thanks to two scientists who contrasted greatly in their work and their personalities and who are forgotten to all except the most dedicated historians of science. The first is the French scientist Henri-Victor Regnault, a man so famous in his time that Gustave Eiffel chose him as one of seventy-two French scientists memorialized on the Eiffel Tower. The patient and careful Regnault spent nearly thirty years documenting—in three volumes containing thirty-two hundred pages—the properties of steam. He devised clever gadgets and gauges to measure, at temperatures from -29°F to 446°F, the energy contained in steam, its volume, and the energy required at these temperatures to turn steam into liquid water and vice versa. On his death, the scientific community celebrated Regnault's "unbending resolution, to place the modern physicist and chemist in possession of an invaluable collection of constants, which are in daily use, not only in the laboratory of research, but for a large variety of industrial purposes."[15] Regnault's data on steam were essential to Parsons's design of a steam turbine, but to use the data fully, he needed to take advantage of theoretical advances in the science of steam.

Parsons's second scientific resource was the theoretical work of William John Macquorn Rankine, a Scottish scientist and a founder of thermodynamics. In contrast to Regnault, Rankine was anything but diligent, quiet, careful, and conscientious. He was a born performer, as likely to sing at the British Association, the most important scientific meeting in the United Kingdom, as to deliver a paper. Ten years or so after Regnault began his work, scientific papers gushed from Rankine's desk—a "veritable tsunami," said one historian of science.[16] These laid the foundation of thermodynamics, although they often built off his idiosyncratic, now forgotten, "hypothesis of molecular vortices," which we might today consider an early atomic theory. Yet his work showed engineers how to calculate using thermodynamic properties. His textbooks were the standard for university-trained engineers through the second half of the nineteenth century and even into the twentieth. Of importance to Parsons's steam turbine was an 1870 paper by Rankine on a phenomenon much simpler than his complex theories of vortices: how to calculate the velocity of steam from a nozzle using Regnault's data. Parsons knew that when combined with Regnault's steam table, a "successful steam turbine ought to be capable of construction."[17]

Parsons's insight was to use these data to design a device that slowed the rotation and the speed of the steam when passed through a set of bladed wheels along a shaft.[18] With Regnault's steam data and Rankine's method of calculation as guides (rock-solid rules of thumb!), Parsons could estimate the velocity of the steam at every step. From this, he learned that to have a modest pressure drop through the turbine, he needed thirty bladed wheels, each like a paddle wheel, arranged on a long shaft inside a cylinder. Parsons blasted steam into one end of the cylinder at about three times the atmospheric pressure. Inside, it struck the first bladed wheel, thus spinning it and rotating the shaft. As the steam passed through the blades, its pressure dropped only slightly, so its velocity was about 275 miles per hour. The steam then passed through a fixed

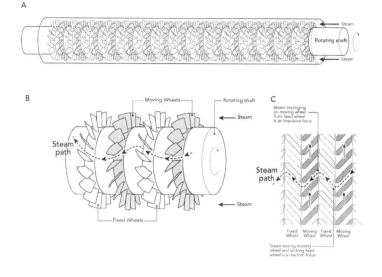

To create the first successful steam turbine, Charles Parsons arranged thirty bladed wheels along a shaft (a.). At each wheel, the steam expands and spins the turbine's shaft. By dividing this expansion over thirty wheels, Parsons avoided the tremendous speeds that destroyed devices like the aeolipile and Branca's paddle wheel. Of the thirty wheels in Parsons's turbine, fifteen were attached to the shaft and rotated with it, and fifteen of the wheels were fixed—they were attached to the turbine casing, not shown here. Steam struck the blade of a moving wheel (b.) and rotated the shaft, then passed through the fixed wheels, which redirected the steam to strike the next moving wheel. When the steam impinged on the blades of a moving wheel, as shown in (c.), it imparted an impulsive force (as in Branca's steam-drive paddle wheel) and when the steam exited a moving wheel and struck a fixed wheel, a reaction force (as in the aeolipile) propelled the moving wheel.

wheel—attached to the cylindrical casing so it could not rotate—with blades angled in the opposite direction of those on the first wheel, and its pressure dropped again by 5 percent or so. This fixed wheel redirected the steam onto the third wheel, which had blades in the same direction as the first, but the angle had changed to increase the space between the blades because the steam had expanded—how large this spacing had to be he estimated from Regnault's and Rankine's work. The steam continued its passage through the remaining fixed and rotating bladed wheels, each designed to accommodate the expansion of the steam and to keep each wheel rotating the shaft at the same rate as the others. If the steam within the turbine were released to the open atmosphere as others had attempted for centuries, it would have destroyed itself with a furious rotation of 300,000 rotations per minute. But within Parsons's carefully, scientifically estimated series of steps, the force was tamed, in his prototype, to a manageable 18,000 rotations per minute. The result was the world's first working turbine.

Highlighting Parsons's use of scientific knowledge is not to imply that he merely needed to implement the scientific knowledge and all was done. No particular scientific data or theory revealed how to make a turbine. There is nothing inherent in the documented properties of steam, collected by the application of the scientific method, that forms the design of a working turbine, waiting to be discovered within the numbers, like Michelangelo's *David* in a slab of marble. The two hundred or so engineers who had made their own attempts at turning steam power directly into circular motion were no less engineers for having not made use of Regnault's data and Rankine's method of calculation, but they had failed to recognize one of the many opportunities engineers have to use science as a powerful tool. Parsons knew how to use science in his engineering because he was among the first class of university-trained engineers, who were steeped in science and mathematics—the way we still train engineers today.

The astronomical number of dimensions and configurations of rotors and blades and every other design variable in the turbine was vast, and without a way to narrow down the possibilities, Parsons's predecessors were lost. For Parsons, scientific knowledge helped rule out what wouldn't work, narrow the possibilities for what would, and shorten the path to a solution. In this case, the vast number of choices were, said Parsons, tamed by "accurate observation and tabulation of facts."[19] His estimate of the steam velocity in his turbine assured him the machine *could* work and guided his initial choice for the number and diameters of the rotors and size of the blades. Hence Parsons's telling statement, "the practical development of this engine was thus commenced chiefly on the basis of the data of physicists."[20] The reliable heuristics from scientific knowledge eliminated fruitless paths and set him on a path that would converge via trial and error—a decade for Parsons—into a successful marine turbine. And that is the relationship of science and engineering: scientific practice and knowledge offer engineers gold-plated, grade A, supremo rules of thumb, rules that work better than those extracted merely from observation or long periods of trial and error, rules that reduce the time to a solution.

At the end of his ten years of careful development, Parsons was stunned that few were interested in his turbine. Three months after his presentation failed to kindle interest at the Royal Institution of Naval Architects, the British government prepared a nautical spectacle to celebrate the sixtieth year of Queen Victoria's reign. They crammed 165 battleships, destroyers, cruisers, gunboats, and torpedo boats into the Solent, a five-mile-long, two-mile-wide strait that separates the Isle of Wight from the mainland of England. The ships, operated by thirty-eight thousand officers and crew, were a tangible display of the power of the British Empire. The Royal Navy was, declared one newspaper, the "living bulwark of our islands" that safeguards "the heart of the Empire."[21] This extravaganza was attended by the wealthy and powerful, who arrived in specially scheduled tours aboard Cunard's giant RMS *Campania* and the White Star's SS *Teutonic*, a predecessor of the *Titanic*. From the decks, passengers admired the martial pageantry on board the warships—sailors in dress blue jackets and scarlet-coated marines—as strains of a brass band wafted from the ships, punctuated by the occasional service bugle calls. With the sound of the signal gun announcing the Prince of Wales, civilian guests on military vessels scurried to their cabins as sailors and officers rushed to their posts on the bridge, turret top, and gunwale.

As the *Victoria and Albert*, carrying the Prince of Wales, started its inspection of the twenty-five miles of the fleet, Parsons's *Turbinia* smoothly accelerated to join the parade and stopped by the *Teutonic* to pick up Thomas Ismay, the founder of White Star, and his son Bruce, who would later survive the sinking of the *Titanic*. Both wanted to ride on Parsons's new "scorcher," as more enthusiastic rumors had described the *Turbinia*.[22] As they boarded, the Ismays swapped their top hats and velvet-collared overcoats for oilskin hoods and overalls, trading the opulence of the *Teutonic* for the spartan fittings of the *Turbinia*—an austere deck of black iron plates, its only shelter a curved partition with a semicircle of windows for the pilot. With not a deck chair in sight, the Ismays stood forward of this shelter and held a thin railing. The *Turbinia* hid behind the stern of the HMS *Powerful*, the largest cruiser in the world, a ship some three hundred times the *Turbinia*'s size.

When the royal yacht passed the *Powerful*, the *Turbinia* made its move. Parsons revved his turbine engine to cut across the line of battleships and trail the royal yacht. Patrol boats set upon the *Turbinia* to force it to the section of the Solent reserved for spectator ships, but the *Turbinia*'s lookout alerted the captain, who yanked the marine telegraph to full speed. In the engine room, Parsons fired up his turbine. A mighty, low-pitched rushing sound from the ship's funnel echoed across the Solent as a long flame belched from the ship's funnel, blistering the paint and raining hot cinders onto the crew. In twenty-eight seconds, the ship's turbine turned at full force with 2,100 horsepower, making the *Turbinia* four times more powerful than any other vessel its size. The ship's bow rose as a greenish-blue sheet of water crashed onto the deck and cascaded over the Ismays, now doubled over to withstand the wind. The ship's smooth wake turned into a white mass of seething foam as the tiny ship zoomed to an astonishing thirty-five knots. The ship, said one observer, "tore through the sea like a thing bewitched."[23] A patrol boat charged toward the *Turbinia* to head it off, but the *Turbinia* darted, coming so close that the patrol boat's lieutenant, assuming a collision inevitable, unbuckled his sword so he could swim from the wreckage. The other patrol boats pursued, but all were at least fifteen knots slower than the *Turbinia*. Eventually a patrol boat managed to block their path, and the *Turbinia* slowed, returned to the *Teutonic*, and dropped off the impressed Ismays.

Parsons gathered tremendous publicity from his stunt. The *Times* called it "adventurous and lawless" and "an effective advertisement," although the stunt was not so "lawless," having been pre-approved by the Admiralty; moments before the *Turbinia*'s jaunt, a patrol boat had dropped off a message from an admiral allowing the demonstration of its great speed.[24] Ten years after the *Turbinia*'s reckless display at the naval review, Parsons's turbine dominated marine propulsion. The British Navy's tests of the turbines in destroyers and the *Amethyst* cruiser, a 3,000-ton ship, went so well that they adopted it for the mighty 18,410-ton *Dreadnought* and all ships that followed. Parsons's turbine also beat out the reciprocating steam engine for fast passenger service, with White Star and Cunard adopting it; White Star most often used multiple of these turbines to augment their reciprocating engines as on the *Titanic*, but the two Cunard ships RMS *Lusitania* and RMS *Mauretania*, then unprecedented in size and power, relied only on Parsons's engine.

Parsons's turbine still affects our lives. For marine propulsion, it lives on in nuclear submarines, where it requires no fossil fuel exhaust, operates quietly, and enjoys a practically unlimited fuel supply because it only needs water heated by a nuclear reactor. More profoundly, Parsons's turbine still enables the daily lives of nearly every human on the globe as its descendants continue to generate the world's electricity with steam power. Fossil fuels, mostly natural gas and coal, or a nuclear reaction boil water to create steam that blasts through a turbine to generate the electricity in your home. Parsons's steam engine perfectly illustrates the relationship of science to engineering: the former generates the best rules of thumb, and the latter applies them to change the world. His work also demonstrates that to call engineering "applied science" hides the creativity of engineering—it obscures Parsons's innovative work by suggesting that once scientists described the properties of steam, the heavy lifting was done. And it is the fuzziest of thinking, conflating the tool with the method. It's akin to saying that carpentry is "applied hammering," that composing music is "applied pitch," or that writing a book is "applied lettering." In the next chapter, we tackle another myth about the nature of engineering: that at its root, it is simply mathematics—applied mathematics, of course.

7
MATHEMATICS

How Engineers Offend Mathematicians
to Predict the Future

The iconic Chicago skyline stuns with its many famous skyscrapers, the most prominent being the Willis Tower, one of the tallest buildings in the world. While it's dizzying to look at from the ground, spectacular when viewed from a jet landing at nearby O'Hare, or delightful as an observation deck, most people simply think of it as just another building, only larger. Yet to an engineer, a skyscraper like the Willis Tower is something completely different: it is a delicate balance between cost and wind![1] As a building gets taller, it becomes more flexible in the wind. To see this, think of a skyscraper as a diving board. If you stand at the end of the diving board, your weight, like the wind on a building, causes it to flex. If the board is made longer, it will flex more under your weight, unless it is made thicker. A skyscraper is just like this: the taller it is, the stronger it has to be to resist the wind. And here is where money enters. As a building gets taller, the amount of material needed increases more quickly than its height, and thus the costs escalate—at least by conventional methods. So to design a skyscraper, an engineer must know the minimal amount of material necessary or, conversely, the strongest wind that will strike the building. To do this, engineers use this rule of thumb: design a building so as to be undamaged by the strongest wind likely to occur once every fifty years and that will sustain only minor damage from the strongest wind likely to occur once every one hundred years.

You can observe the results of this guideline yourself. Stand still on the top floor of the Willis Tower on a windy day, close your eyes, and you'll feel it sway. Fill a bathroom sink with water, and you'll see it gently slosh back and forth. The tower, like all skyscrapers, is built with just enough steel to resist the wind, just enough to test fate at the hands of unpredictable forces of nature yet come out on top. To find the speed of those winds, an engineer flips open a book from the American Society of Civil Engineers titled *Minimum Design Loads and Associated Criteria for Buildings and Other Structures* and examines a map that plots region by region these "fifty-year" and "hundred-year winds," the engineering shorthand for these maximum winds. The rule's simplicity, even simplemindedness, its ease of use—merely reading a map!— obscures a mystery at its heart, a single word glossed over: it uses the *likely* winds. In other words, engineers need only predict the future.

To create these maps of likely maximum winds at first seems simple: study the record of the last one hundred years of wind data, perhaps daily measures of the maximum wind speed, then use the maximum wind that occurred as the hundred-year wind and the maximum wind of the last fifty years as the fifty-year wind. Yet as in all engineering problems, incomplete information and uncertainty intrude. The Willis Tower was designed based on only forty years' worth of total data—a typical amount of data used even today to build a skyscraper—with Chicago's downtown wind speed measurements made at thirty feet above the ground and upper air

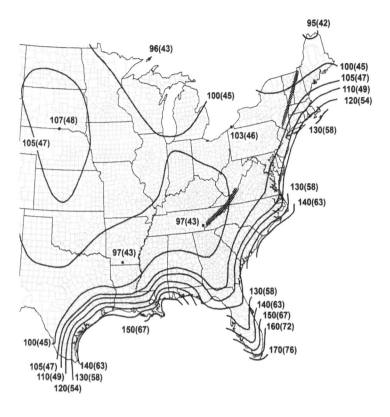

95(42)
96(43)
100(45)
105(47)
110(49)
120(54)
100(45)
107(48)
105(47)
103(46)
130(58)
130(58)
140(63)
97(43)
97(43)
130(58)
140(63)
150(67)
160(72)
150(67)
170(76)
100(45)
105(47) 140(63)
110(49) 130(58)
120(54)

The labels on contours on this map are the fifty-year wind speeds in miles per hour (meters per second).

measurements taken outside the Loop. To overcome this constraint of having only a limited set of wind data—how engineers would love to have hundreds of years of data!—engineers developed a rule of thumb for their rule of thumb that can answer this question: how to use a sparse set of data, perhaps thirty or forty years' worth, to predict their fifty- and hundred-year winds.

Rules of thumb that predict the future are classified as "risk assessment," a dull term for a methodology that predicts catastrophic events, a discipline that estimates the occurrence of devastating floods, record snowfalls, and gale-force winds, that assesses the number of deaths caused by a particular highway design, or that quantifies how likely spent nuclear material is to seep from its storage container. Risk assessment rules highlight the relationship of mathematics to engineering. Like science, mathematics empowers engineers with superb rules of thumb, but with a twist: mathematics enables engineers to predict the future. It brings to the table probability and statistics to create engineering rules of thumb at their most audacious. Risk assessment methods are rules of thumb on steroids. A mathematically sophisticated framework developed over centuries is captured succinctly in the rules of thumbs used by engineers to design safe dams, buildings, highways, and nuclear storage with limited information. The rule of thumb to predict "likely" fifty- and hundred-year winds emerges from a thousand years of thought that began with a move away from predicting the future by fortune-telling, omens, and astrology to assessing risk by synthesizing observations and data into a best guess.

The earliest risk assessors lived in the Tigris-Euphrates valley in 3200 BCE. They were a secretive cult called the *āšipu*, sometimes translated as "conjurer," with connotations of "priest," "exorcist," and "physician," but the title is really untranslatable. Another of their roles would be considered "soothsayer." To foretell someone's future, a member of the āšipu conducted a detailed examination. They felt the skin to assess its temperature and studied the arrangement and color of blood vessels near the skin's surface. For pregnant women, these blood vessels foretold the severity of labor pains and the gender of the child, for the ill the duration and even fatality of their illness.

Over centuries, the work of the āšipu evolved into an early form of risk management—no different in principle than that of engineers predicting winds—as they moved from merely telling fortunes to clearly defining possible futures, collecting data about them, then, after assessing their likelihood, assigning a plus or minus to each outcome. As careful planners, they became valued members of royal missions. When Esarhaddon, a powerful king of the Neo-Assyrian

Empire, invaded Egypt, he brought along nine āšipu to assess battle plans and possible outcomes. Still, because Esarhaddon believed that the gods created the universe so the wise could read its signs, in tow were seven astrologers, five diviners, six lamenters, and three bird-watchers. Yet the seed of a rational method to estimate risk had been planted: the crucial idea of collecting data from observations. This is an idea we see when today's engineers search for clues in thirty or forty years of wind data to forecast wind speeds. The breakthrough that turbocharged this kind of careful data gathering and pushed astrology, divination, and the reading of tea leaves to the periphery was the development of mathematical probability—assigning a numerical value to the likelihood of an event occurring.

A refined idea of probability grew from a chance meeting of France's greatest seventeenth-century philosopher and an over-whelmed courtier at the court of Louis XIV. The king's courtier, Antoine Gombaud, chevalier de Méré, so charmed the glitterati of Paris that every salon in the city demanded his presence to the point of his exhaustion. To be restored by quiet contemplation, Méré traveled every two months to his hometown of Poitou to sit by the town's picturesque Atlantic coastline. On one of his many visits, Méré was accompanied by his friend the Duke of Roannez. The duke, a gifted amateur mathematician, invited a guest to share their coach. "In order to relieve tedium," Méré later wrote, "he had provided a middle-aged man, who then was very little known, but who later certainly has made people talk about him. He was a great mathematician who knew nothing but that. These sciences give little sociable pleasure, and this man, who had neither taste nor sentiment, could not refrain from mingling into all we said, but he almost always surprised us and often made us laugh."[2]

The mathematician, Blaise Pascal, also amused Méré with his odd behavior: Pascal would draw from his pocket a packet of strips of paper, remove one, and write an observation or a thought, then tuck it away. These thoughts, despite Méré's misgivings about the "sociable pleasure" of mathematics, drew his attention since he had the opportunity to pepper Pascal with questions about chance.

Chance, fate, and fortune fascinated all thinkers in this era of Enlightenment, which wrestled with human free will versus the foreknowledge of all events by God. They ardently read *The Consolation of Philosophy* by Boethius, a bestseller since its publication in the sixth century, a book so much of the age that Queen Elizabeth I famously translated it near the end of the sixteenth century. The autobiographical book, written while Boethius was imprisoned in an Ostrogoth jail because of suspected sympathies for the Byzantine Empire, introduces the notion of a wheel of fortune to explain chance. In the book, Boethius encounters, in his jail cell, Lady Philosophy, who explains that fortune is like a wheel that turns "so that what is low is raised high and what is up is brought down. You ascend? Fine! But you must acknowledge that it can't be wrong for you to have to descend again."[3] Life can turn on a dime, and the most powerful king can find himself destitute in a day. So in their discussion of chance, Méré posed to Pascal a conundrum from the late fifteenth century about an interrupted game of chance.

In the game, called "points," two players each placed a wager, then competed to earn points. The first player to reach, say, four points won the wager. The competition was simply a coin flip in which each player gained a point when the flip either came up heads or tails. In our age, saturated in statistical and probabilistic knowledge, this game perhaps seems dull or at best arbitrary, but in an era wrestling with the meaning of fate like Boethius had before, the game fascinated, especially the problem of an interrupted game. How, then, should the wager be awarded if the game was stopped before either player reached four points? Most accepted the answer given by the sixteenth-century Italian mathematician Niccolò Fontana Tartaglia: "the resolution of such a question is judicial rather than mathematical, so that in whatever way the division is made there will be cause for litigation."[4] Méré received no quick answer from Pascal, but he planted a seed that grew into the cornerstone of our modern uses of probability.

Fascinated by the problem, Pascal wrote to mathematician Pierre de Fermat, famous to us for his "last theorem." In the margins of a book on mathematics, he wrote an addendum next to an ancient theorem and insisted, "I have discovered a truly marvelous proof of this, which this margin is too narrow to contain," yet it eluded proof for 358 years.[5] In 1654, Pascal wrote to Fermat: "Monsieur le Chevalier de Méré is very bright, but he is not a mathematician, and that, as you know, is a very grave defect."[6] In their letters that followed, Pascal and Fermat considered a game of points where one player had two points and the other had one. To win, then, player A must earn two more points, and player B must win three more points. Pascal and Fermat first enumerated the possible outcomes aided by a formula derived by Pascal. "The labor of the combination is excessive," wrote Pascal. "I have found a shortcut and indeed another method which is much quicker and neater."[7] For this interrupted game, if player A with two points won when heads came up (H), and player B with one point won if tails came up (T), then the game needed four more tosses at most. Using Pascal's formula (and common sense), they enumerated all the possible outcomes for the next four tosses.

Out of these sixteen futures, player A won eleven times, while player B won five times. In a key step, Pascal used this analysis to answer the question of how to split the wager for the interrupted game: player A should get eleven-sixteenths (68.75 percent) and player B five-sixteenths (31.25 percent).

Again, to us, this kind of solution, though a bit intensive for a coin-flipping game, seems simple if not intuitive. We live in a world

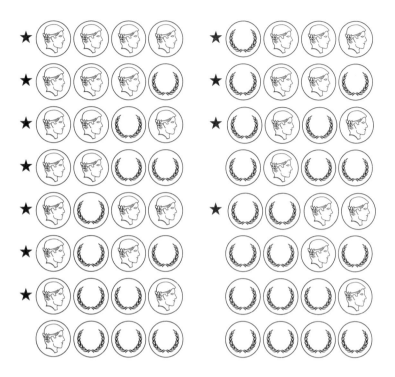

Pascal and Fermat enumerated all the possible outcomes of the next four flips of a coin for a game of points where player A has already earned two heads and player B has earned one tail. The winner was the person who got four points. In the figure, the starred sequences indicate when player A wins.

run by predictive risk analysis evaluating inclement weather, diseases, market speculation, election outcomes—an entire modern society held up by decision-making rules of thumb empowered by the most accurate knowledge we can generate (though never nearly as accurate as we would like) about whether and what bad things will happen in the future. Pascal was a part of creating that world. With this revolutionary idea, he placed the cornerstone of risk analysis: to use probability to assign numerical values to a range of possible outcomes—a refinement of the pluses and minuses assigned to different futures by the āšipu. But so far, Pascal's work was just a matter of calculation.

His solution to the interrupted game of points didn't involve any problem-solving or decision-making or creating something in the physical world, which would turn it into a rule of thumb used by engineers. Pascal later extended the idea beyond puzzlement about a game of chance to generate his famous rule of thumb regarding another key Enlightenment question with the small slips of paper that amused Méré on their trip to Poitou.

Pascal gathered his scratch paper notes to create his great work *Pensées* ("Thoughts"), in which he proposed his famous "wager."[8] He wrote, "Either God is or he is not. But to which view shall we incline?" He assigned probabilities to God's existence or nonexistence: "a coin toss is being spun which will come down heads or tails." It is, he implied, a fifty-fifty shot, although he later admitted that the probability of God's existence might be smaller. He assigned utility and value to the outcome of believing or not believing—a yet more sophisticated version of the plus-and-minus system of the āšipu. Following the prevailing Christian beliefs of seventeenth-century France, to believe in an existing God granted an eternity in paradise after death, while disbelief damned one to hell. Belief in a nonexistent God, however, only compelled one to live according to religious standards based on a misconception with no eternal reward. Disbelief in a nonexistent God gave one more freedom in life while being left to the same fate after death as the believer. Following the āšipu's model, belief in an existing God provided essentially ultimate, infinite benefit, and disbelief in God provided infinite loss. If God didn't exist, belief provided finite loss, while disbelief provided finite benefit. Given the stats, the best one can get from belief is eternal bliss, whereas the best disbelief offers is a marginally better life on earth. The worst one suffers from belief, on the other hand, is a life inconvenienced by religion, while with disbelief comes the possibility of eternal damnation. So Pascal concluded that a risk analysis compels belief in God. Although the theology, even logic, of Pascal's wager has been debated for centuries, he shows us a turning point in probability theory: the idea that the mathematics of games of chance can be applied to a wide swath of life, that probabilities can be developed into an art of conjecture, a way to think about future events with some enumerated degree of confidence that we can use to make informed decisions to mitigate risks and solve problems.

In the centuries that followed, the application of Pascal's probabilistic analysis moved from answering deep, theological questions to the most earthbound of things. Statistics on poverty, health, mortality, and crime were collected by public agencies, then, by the middle of the nineteenth century, combed through by statisticians to detect patterns like stable crime rates or the changes in mortality rates from year to year. As the century progressed, they used these statistical averages to document deviations from typical patterns that gave clues to the causes of crime, poverty, and death. Statistical analysis pinpointed the source of London's cholera outbreak in 1854, and Florence Nightingale's precise statistics on causes of death in the Crimean War led to better sanitation in hospitals.

In short, the Enlightenment and Industrial Revolution that followed turned mathematics into a supercharging force for every kind of engineering with a new, implicit, and nearly omnipotent rule of thumb: quantify or express all variables as numbers. Once again, to us, this seems obvious. Even the medieval masons' proportional rule used numbers and math to figure the size of a supporting wall, and that was hundreds of years before Pascal's *Pensées*. Of course, math has always been an inextricable part of engineering, but with this era came the practice of applying numbers to every aspect of a technical problem, whether directly observable or only probabilistic. And with science experiencing a similar breakthrough, engineering had in mathematics a tool for combining scientific knowledge with workarounds for uncertainty.

Although these statistical methods quantified the world, this notion of averages, so useful for finding and making use of regularities and patterns, does not solve the engineer's problem of extracting from thirty or forty years of daily wind data the likely maximum winds that will occur every fifty to one hundred years. The average wind over fifty years is of no use to an engineer designing a tall building: they need an extreme, an outlier. Engineers must know about the *least* likely events, about extremes like devastating floods, violent snowstorms, and gale-force winds. But the principles of predicting extreme events didn't come from grave issues of life and death like deadly diseases or collapsing buildings. Instead, an industrial physicist needed to know whether a strand of yarn was likely to break while weaving.

In the early 1920s, the British Cotton Industry Research Association awarded twenty-three-year-old Leonard Tippett a scholarship to study statistics at the University College, London. Their hope was that Tippett, freshly graduated with a degree in physics from the Royal College of London, could improve the weaving of cloth, which, on an industrial scale, is a dazzlingly complex process. To weave a piece of cloth, hundreds of strands of yarn are pulled between two rollers. Through these strands a shuttle threads back and forth two or three miles of weft yarn contained in one long woven strand. If one thread per yard of cloth breaks during the process, the weaving becomes, as Tippett noted, "impossibly difficult."[9] So although the statistics of the day could calculate the average strength of any piece of yarn, Tippett needed to know the weakness of the weakest fiber in a bundle of strands ready to be woven into cloth. "In many things of ordinary experience," he reasoned, "the occasional extreme abnormality is far more apparent than a large number of small abnormalities." A train that is thirty minutes late on 5 percent of its journeys will be perceived by riders as having a worse reputation than one that is five minutes late on 30 percent of its journeys. Few care about the average time a train is late: it is the distribution of these times that interests a passenger. Tippett needed a way to measure the strength of a sample of the yarns, then use that information to predict the likelihood that a weak strand would be in that batch. To solve this problem, he partnered with the greatest statistical genius of the day, Ronald Fisher.

Fisher, according to the great science writer Stephen Jay Gould, hit so many "home runs" with his newly developed statistical methods in the first quarter of the twentieth century that he should be called the "Babe Ruth of Statistics."[10] The nickname hasn't stuck with the Edwardian-era Brit, but his innovative statistical techniques shape the work of every statistician today. He initially studied agricultural genetics, where he developed his deepest statistical insights over fifteen years at the Rothamsted Experimental Station in Harpenden, England, about twenty-five miles north of London. At Rothamsted, Fisher established himself as the "go-to" statistician, and soon laboratories and organizations from all over the world wrote to him for advice and sent their researchers to study with him. These "voluntary workers," as they were called at Rothamsted, returned to their home countries and institutions with Fisherian statistical methods and Fisherian experimental design. Tippett was among the first in this wave of voluntary workers.

The precise question that Tippett put to Fisher was this: How likely was a particular amount of weight to break one of the hundreds of pieces of warp yarn feeding the loom? He could not, of course, unravel the yarn wrapped around the warp roller and test each section of yarn for strength: he would have no yarn left with which to weave! He needed to sample the strength of a few of the yarn bundles wrapped on the warp roller, then predict the likelihood that *all* the yarn would be strong enough. He might sample, say, thirteen sections of the bundles of yarn wrapped around the roller: cutting each into sections, testing the strength by adding weights until they broke, then recording *only* the weakest piece of yarn in the bundle, namely, the one that broke with the smallest amount of weight. With the criterion being that at least 99 percent of the yarn strands must be likely to be able to hold a seven-gram weight without breaking in order to be fit for weaving, he might find this:

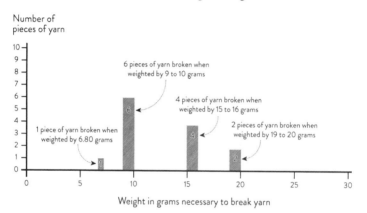

Number of pieces of yarn

6 pieces of yarn broken when weighted by 9 to 10 grams

4 pieces of yarn broken when weighted by 15 to 16 grams

1 piece of yarn broken when weighted by 6.80 grams

2 pieces of yarn broken when weighted by 19 to 20 grams

Weight in grams necessary to break yarn

To predict whether the yarn in the bundles on a warp beam will be strong enough, one can sample thirteen of the bundles from the hundred or so used in the loom: cut each bundle into short sections, and find its breaking strength by hanging weights from it. Shown above is the *least* amount of weight needed to break one of the sections of yarn. For one of the bundles, the weakest section broke with a weight of just under seven grams; for six of the bundles, the breaking strength of the weakest yarn was nine to ten grams; for four bundles, fifteen to sixteen grams; and for two bundles, nineteen to twenty grams. These results can be used to predict the behavior of *all* the bundles.

One of the pieces broke with a weight of less than seven grams; six of them required nine to ten grams; four, fifteen to sixteen grams; and two strong pieces needed nineteen to twenty grams to break. To use these thirteen samples to predict the behavior of all the bundles took the statistical genius of Fisher.

Fisher demonstrated mathematically that the distribution of the strengths among the bundles should follow this shape:

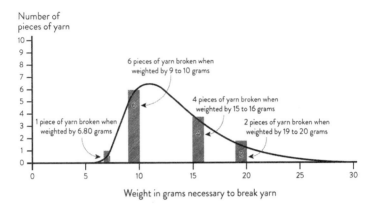

The line superimposed on the breaking strength of the weakest section of each of the thirteen bundles is the *probability density function*. With it, you can estimate the probability of a
range of breaking strengths: the area equals the probability.

This curve is called the probability distribution function, a
mouthful but simple to understand. The probability of a piece of
yarn having a breaking strength between two values is proportional
to the area under the curve between those two points—the total area
under the curve is one, because a piece of yarn will eventually break.
So if we look at the area covering 99 percent of the total area, we can
see whether the value of 99 percent of the yarn would require more
than seven grams to break.

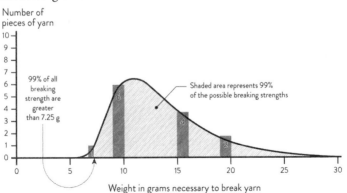

The shaded area covers 99 percent of the area under the curve. This means that ninety-
nine of the bundles have a breaking strength greater than 7.25 grams.

From this graph, we see that 99 percent of the bundles have a breaking strength greater than 7.25 grams, so the yarn on the roller would likely be strong enough.

Consider now a set of samples of eighteen bundles from a different lot of yarn. It might reveal this:

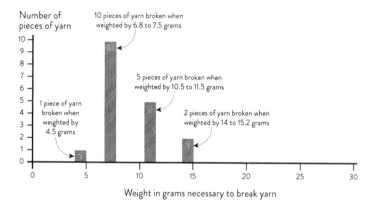

From a different warp beam, one might sample eighteen bundles and see a different distribution of breaking strengths. The weakest section of one bundle was 4.5 grams; for ten of them, 6.8 to 7.5 grams; for five bundles, 10.5 to 11.5 grams; and for two of the bundles, 14 to 15.2 grams.

Fisher's curve can be fit to this (the same shape but squeezed a bit):

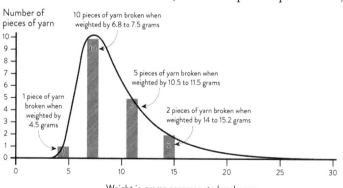

The results from these eighteen bundles can be fit with a curve of the same shape as for the earlier set of bundles but compressed a bit.

For this sample, we see that less than 99 percent of the fibers have a strength above seven grams:

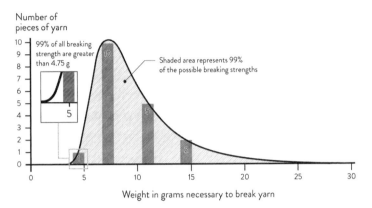

For this set of bundles, 99 percent of them have a breaking strength greater than 4.75 grams. Although not easy to tell from the graph, only about three-quarters of the bundles would likely be strong enough: too low to risk using this batch of yarn for weaving.

Tippett would conclude that it was not worthwhile to risk using this yarn to weave cloth. Although difficult to tell from this graph, only about three-quarters of this yarn is likely to be strong enough.

The strength of a piece of yarn seems far afield from the likelihood of the occurrence of gale-force winds or how to take thirty years of daily wind observations and extract a likely maximum for every fifty or hundred years, yet they are intimately connected. The mathematical technique developed by Fisher and Tippett could be used for *any* extreme event.

The power of these kinds of analyses has been a priceless tool for every kind of engineering since the early 1900s, but the era in which they emerged made them convenient tools for the most willful perversion and misuse of science in Fisher and Tippett's era: racial eugenics.[11] Fisher's personal interest in statistics was sparked by a desire to improve the world by genetic selection, encouraging childbearing by those whom he and his colleagues considered of the best genetic stock and denying it to those who didn't meet his standard.[12] "The policy of eugenic sterilization," said Fisher, "affords the only practicable remedy for some of the saddest afflictions to which mankind is subject."[13] His belief in eugenics was sadly common at the time, with the British eugenics movement of the early twentieth century including influential figures like H. G. Wells, George Bernard Shaw, John Maynard Keynes, and Winston Churchill. But the use of statistics to "prove" the necessity or efficacy of eugenics in any kind of rational sense was a corruption of statistical analysis as a problem-solving tool, using it to confirm preconceived animus against those without Western European heritage by selecting convenient data while categorically excluding contradictory information or historical forces. Yet statistics was still a force for exposing trends and facts, and there was one statistician who used it against the emerging Nazi Party.

Emil Gumbel, a contemporary of Fisher and a German Jew, spent his professional life studying the statistics of extreme events, but he also spent a lifetime living through extreme events. During the Weimar Republic, at the time Tippett and Fisher developed their theory of calculating extreme values, Gumbel watched with horror as the rise of Hitler's Nazis in Bavaria became a palpable threat to Germany's future. They used political murder as a tool to consolidate their power while being protected by a political and judicial majority that, fearing the rise of socialist sentiment in an impoverished Germany, was willing to give a pass to fascist thugs who terrorized their political opponents. To expose the tacit political alliance and hypocrisy of German judges, Gumbel applied statistics to the verdicts and sentences of the trials to prove that the Nazi murderers were let off lightly. His statistical analysis of the published budget of the Black Reichswehr, the paramilitary group formed in violation of Treaty of Versailles prohibitions, revealed the hidden allocations for illegal rearmament and support of terrorist organizations. From the death rates of German soldiers in World War I, he highlighted the impact of starvation rations, which undercut the Nazi propaganda that claimed internal betrayal by socialists and German Jews lost the war for Germany and led Gumbel to conclude that the heroes were not soldiers or generals but those who opposed the war. He suggested that instead of erecting monuments glorifying Germany's valor during the Great War, the symbol memorializing the war should be the turnip, as a reminder of the starvation and suffering of the German people in the name of the war. Angered by his blistering critiques, the Nazis designated him a traitor, stripped him of his citizenship, and issued an arrest warrant, which would lead to certain execution. He escaped to the United States, where, shielded from the horrors of the Nazi regime, he studied extreme events in the abstract.

In the final twenty-six years of his life, Gumbel applied Tippett and Fisher's statistics of extreme events to many problems in physics, engineering, industry, and the social sciences. He was driven by his desires, as in Germany, to use his statistics to benefit humankind. Although the objective world of statistics was principally uninterested in this, his mathematical work always had a human application. Using data from both the Rhône and Mississippi rivers, he demonstrated to engineers how to calculate the typical return period of floods so they could size dams, levees, and reservoirs. He returned to problems like that of the breaking strength of yarns and showed how engineers could sample electronic components like capacitors and predict how often they would fail. For the manufacture of light bulbs, he demonstrated how to sample a few, measure when they burned out, then use that data to estimate the variation in the lifetime of a light bulb. He used extreme value theory to find the necessary time to kill bacteria by a disinfectant, which, like Tippett's yarn strength, couldn't benefit from knowing the average time needed—all the bacteria needs to be gone. He even applied it to the stock market, not to predict the future—"no statistical method can forecast the value of a stock at a certain day," he said—but to decide whether the Dow Jones Industrial Average deviated from that in previous years.[14] And in one fascinating application, from a man whose life span was often uncertain, being under the threat of death by political murder, he applied his statistics to estimate the extremes in the span of human lives and the question whether there is an upper limit to how long a human being can live. Gumbel's extensive, comprehensive, and careful work laid the basis for one more application: calculating the hundred-year wind.

A calculation of the probability of an extreme wind was pioneered in the early 1950s by a U.S. government employee, Herbert Conrad Schlueter Thom. Thom is one of millions who contributed to twentieth-century engineering and science yet whose lives leave little personal trace.[15] Here is what we know. He always signed his scientific papers H. C. S. Thom but was known to all as "Herb." His professional career began in the Second World War, when he analyzed weather to help the Allied Forces' strategic planning. In the late 1940s, he taught agricultural climatology at Iowa State University, but before the end of the decade, he joined the U.S. Weather Bureau, becoming their chief climatologist, and stayed there for decades, a steady presence as the bureau was reorganized and renamed—the National Weather Service, the Environmental Science Services Administration, then the National Oceanic and Atmospheric Administration. The only mention of his death is a single line with his name in a list with fifty-one others in the *Bulletin of the American Meteorological Society* in 2008—no description, no details.[16] Although the record shows no trace of his hair color, his hobbies, or his life, his published work reveals his mind.

Thom loved data—meteorological data. In decades' worth of scientific work, he combed through large data sets to draw conclusions. He studied the amount of rain and crop yields in Iowa to find the minimum rain necessary to prevent drought. He examined daily soil temperatures in April, May, and June for every county in Iowa to estimate the likelihood of a killing freeze. He analyzed the data on the growth of ten thousand corn plants to assess the correct soil conditions to maximize yield. He studied seventy years of measurements of the occurrence of cyclones in the North Atlantic Ocean to predict their frequency of return. He studied data from the "tornado corridor" between Iowa and Kansas to estimate their frequency. And he estimated the chance of cloud cover along the path of an eclipse along the East Coast of North America from Salina Cruz, Mexico, to Gander, Newfoundland. But his most lasting contribution was calculating the probability of extreme winds.

In 1953, Thom studied wind data from Fort Wayne, Indiana, with a list of the yearly maximum wind for the previous thirty-seven years—from 1916 to 1952.[17]

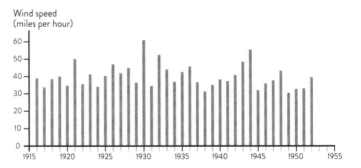

A report of the maximum wind speed every year from 1916 to 1952. Note that 1930 had the fastest wind, just over 60 miles per hour, while 1949 had the lowest wind speed, about 30 miles per hour.

Thom then grouped winds close in speed:

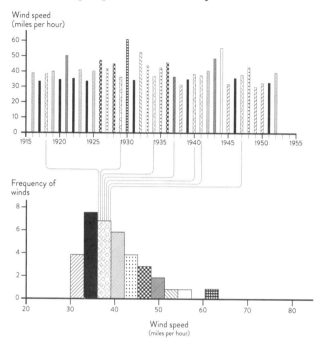

To predict the hundred-year wind, Thom gathered the wind speeds into "bins." The dotted lines indicate the seven winds with speeds between about 35 and 38 miles per hour that compose a bin.

Then, using Gumbel's methods derived from Fisher and Tippett, he fit a curve:

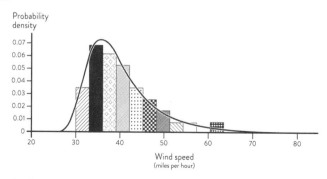

Thom fit a Tippett-Fisher probability density curve to the data.

And he found the speed that 99 percent of the winds were below:

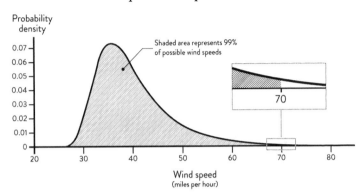

The area under the curve corresponds to the probability for a range. Shown above is the probability of finding 99 percent below a particular speed. As shown in the enlarged section, 99 percent of the winds will be 70 miles per hour or less, so a wind that would appear with a frequency of every one hundred years would have a speed of 71 miles per hour or greater.

He estimated that a wind of 71 miles per hour would occur every one hundred years.

Over the next fifteen years, this vicarious data monger dug into every bit of data the Weather Bureau had collected. He studied large storms, extratropical cyclones, thunderstorms, even tornadoes, until by the late 1960s, he could create a map of the United States showing the hundred-year winds nationwide. The map has been updated and improved often and reprinted in that booklet, *Minimum Design Loads for Building and Other Structures*—a fulfillment of the dream of the āšipu to foretell the future by observing and recording nature. The map is a testament to genius but also to the careful but anonymous work of ordinary men and women.

The rule of thumb embodied by the maximum wind maps of the civil engineers demonstrates that the most powerful application of mathematics to engineering is not mere measurement and quantification but instead the ability of engineers to go far beyond the exactness and certitude of mathematics by using statistical methods.

With these methods, engineers have sized dams, predicted the side effects of drugs, protected vital petroleum pipelines from corrosion, and built safe highways. Yet engineers have used mathematics in a way that mathematicians would condemn as inconsistent and lacking exactness. The method of predicting maximum winds is "anathema to large numbers of statisticians," admits an expert in this field. Many statisticians object to "the extrapolation of statistical models to domains on the fringe of, or beyond, observed data is scientifically unreasonable."[18] And if you examined its mathematical foundations from a philosophical perspective, you would likely become even more alarmed. Philosophers, mathematicians, and statisticians debate whether results like the hundred-year wind are some kind of true probability of what happens in the natural world or whether the only thing that can be said is if one were to *somehow* re-create the weather again and again and again, only one out of a hundred times would this maximum wind, the hundred-year wind, occur, then letting *you* choose at your peril whether this is really the probabilities embedded in the natural world. And indeed, a wind of a never-before-recorded speed could rip through the prairies and city blocks of Chicago, a "black swan" event.

All these objections seem lethal, condemning the methodology of the hundred-year wind, yet they only confirm that the hundred-year wind is an *engineering* rule of thumb—not a mathematical theorem but a context-dependent guide subject to revision like all rules of thumb. The rule for predicting high winds may be changed or scrapped as climate change invalidates the underlying assumption that the global weather patterns over the last hundred years are the same as those of today. Yet for *now*, the time frame paramount to an engineer, this rule of thumb works best to overcome uncertainty about the future. As the first president of the Society for Risk Analysis—the professional organization of engineers and technologists dedicated to estimating risk—noted, "risk assess-ment...must produce these answers because decisions will be made, with or without its input."[19] And doubtless either this rule of thumb or another with just as little philosophical substance will be used to build new buildings, buildings in which tomorrow's philoso-phers, scientists, and mathematicians can debate truth, nature, and certainty.

8

INVENTION

The Myth of the Lone Inventor and
the Race to Light the World

In November 1880, the reading room of the Mercantile Safe Deposit Company, located in the basement of one of the first skyscrapers, glowed with the light of a four-bulb chandelier and six bulbs in fixtures spaced along the walls. An observer characterized this electric light as "very much like that of a first-class oil lamp, steadier than gas, and of a yellow, clear pleasant quality"—nothing like the "ghastly blue" of a "flickering" arc light, nor was there the odor of burning gas; instead, the room's atmosphere "remain[ed] perfectly cool and sweet."[1] His only complaint was that the bulbs flickered slightly with every stroke of the engine that powered the generator. This first commercial installation, a spectacular achievement, featured no bulbs manufactured by Thomas Edison, although he had proudly announced his invention of the light bulb only a few months earlier to great press attention. The bulbs at the Mercantile Company were those of the U.S. Electric Lighting Company, a company driven forward by their irrepressible and energetic engineer, Hiram Maxim. Edison called Maxim's bulb "a clean steal" of his lamp.[2] Yet Maxim had seventeen patents on incandescent lamps, and his company controlled the patents of several other inventors, also contemporary to Edison. Maxim thought of himself as the inventor of the commercial light bulb. "Every time I put up a light," he complained, "a crowd would gather, everyone asking, 'Is it Edison's?'"[3] This so irritated Maxim, who noted that Edison at the time "had never made a lamp," that he considering killing "on the spot" the next person to ask him "Is it Edison's?"

That the first commercialized light bulbs were not Edison's surprises because we love stories of sole inventors whose spark of inspiration revolutionized the world. They give us narratives that are neat, tidy, and digestible but incomplete. These stories hide the engineering method; they bury the creativity of engineers, smooth over struggles, and sanitize choices that reflect cultural norms. Perhaps no story persists more than Edison and his light bulb, yet Edison was the tail end of a long list of light bulb innovators in a process of invention similar to that of the steam turbine in the next century.

In the forty years before Edison's first successful prototype, at least twenty people presented, patented, and demonstrated incandescent lamps—using electricity to heat a filament until it glowed. The first recorded attempt was in 1838 (almost a decade before Edison's birth) by a Belgian inventor whose bulb used a strip of carbon as a filament. A fair assessment of history would call these men inventors of the light bulb comparable to Edison, especially in a world where Edison, the so-called inventor of the incandescent light bulb, was forty years late to the idea of incandescent lighting. But unlike with Edison, we don't remember the names of these men, because most of their bulbs burned for only a few seconds. They had the necessary but thankless job of creating links in a chain of incremental advances that didn't yet produce an applicable or reproducible solution to the problem of darkness, which so far could only be dispelled with fire, until Edison created one of the links that did, transforming from method into narrative. Although Edison and his bulb end that length of the chain of innovators, his link was no more an exercise or example of the engineering method than those that came before; it only overcame a circumstantial threshold of usefulness.

In 1878, Edison focused the energy of his staff at the bustling Menlo Park Research Laboratory on finding a long-lasting filament for the incandescent light bulb. The staff worked to the rhythms of Edison, "the central originating and guiding mind and personality," as one worker noted, describing work there as "a strenuous but joyful life for all physically, mentally, and emotionally."[4] Edison set the tone with long work hours into the night. He often napped on the workbenches in Menlo Park and ate sparingly in increments of small snacks he thought were better for digestion, although for his workers, he had brought in, often at midnight, hamper baskets loaded with hot dinners of meat, vegetables, dessert, and coffee. But when Edison stood, stretched, hitched up his waistband, and sauntered away, all knew that dinner was over and work should resume.

In the late 1870s, Edison and his staff produced bulbs that looked much like a modern bulb: a glass envelope fastened to a wooden base covered with copper strips, and, at its center, a thin, long, delicate spiral of platinum.[5] Yet these bulbs failed. Some yielded light as bright as a small bundle of today's Christmas lights for a few hours, but most burned out quickly. As Edison learned, the temperature for the incandescence of platinum wire was near that of its melting point—any fluctuations in the current and the platinum would melt. Edison and his team tested an astonishing array of materials, by some count sixteen hundred types. They tested metals like platinum, iridium, ruthenium, chromium, aluminum, tungsten, molybdenum, palladium, manganese, and titanium; elements that sometimes behaved like metals, including silicon and boron; then a grab bag of materials—cork, wax, celluloid, and the hair from his employees' beards. After these, his team moved to slivers of wood, broom corn, and paper. Tissue paper covered with lampblack and tar and rolled into a rod glowed astonishingly well and for a good amount of time. Edison refined this idea by "carbonizing" cotton thread, heating it without oxygen until the length of thread was blackened throughout. From this thread, he formed a long filament. On October 21, 1879, a bulb with a filament of this thread, with all the air removed from the glass enclosure, burned for more than half a day. They were approaching the beginning of the commercial light bulb.

Seven months after that bit of carbonized thread showed promise, they tried a piece of bamboo: a six-inch strip burned for three hours and twenty-four minutes at seventy-one candlepower (about the brightness of a standard sixty-watt bulb today). "The best lamp ever yet made," an Edison associate noted, "here from vegetable carbon."[6] From there, Edison's team tested two hundred species of bamboo until they found a variety that was the best for manufacturing carbon filaments, grown near Yawata, Japan, where Edison is still celebrated with a street named "Edison-dori," a bust of Edison in the town center, and, near a shrine, a large monument dedicated to Edison. With his specialized bamboo supply and method of manufacturing in place, Edison was ready to light the world, but Hiram Maxim beat him out of the gate.

Maxim's bulbs, installed at the Equitable Life Building, outclassed Edison's. "They have a rich golden tint, resembling that of a wax taper," said one reporter.[7] Another noted that Maxim "has invented a lamp which surpasses, I believe, even Edison's dreams."[8] When comparing the lamps, reporters noted that Edison's had lower brightness than Maxim's, or, when of the same intensity as Maxim's, they burned out in only a few hours. By Maxim's own estimate, the filaments in his bulbs could last forty days. The dimness and shorter life of Edison's bulbs were the same thing: Edison's bulbs could not tolerate as much current as Maxim's, so if run at the same current, Edison's bulbs would burn out quickly, and to make them last longer, Edison's were run at a lower current and thus were dimmer.

That Maxim could achieve this was unbelievable to Edison's staff—an outraged member of the Menlo Park staff ranted that it must be apparent to "any sane person that" Maxim's bulb must be "but a copy" of Edison's.[9] Surely, thought Edison's employees, only a well-oiled machine like that of Menlo Park could produce a light bulb. Inside Menlo Park, glassblowers, machinists, engineers, chemists, and physicists churned out inventions like appliances on an assembly line, while Maxim's ham-handed U.S. Electric Lighting Company struggled to find enough resources to survive; employees thought it likely to shut at any minute, and even its own president described it as "helpless."[10] Their technical expertise was so low that they could not figure out, as one employee later noted, what "size wire would carry a certain number of lamps without overheating," adding that "a number of mysterious fires about this time were probably the fruits of our ignorance."[11] Compared with Edison's factory-line Menlo Park model, Maxim's method of invention seemed scattershot.

Maxim was the classic American tinkerer, once describing himself as a "chronic inventor." Although self-taught—one biographer describes him as "semiliterate"[12]—over his lifetime, he invented an astonishing array of tools and toys. Maxim developed methods to separate metals from their ores, instruments to measure wind velocities, vacuum cleaners, novelty items that produced "illusionary effects"—a rotating sphere with concave paraboloidal floor, mirrors, and a bicycle track, presumably to create the illusion of riding a bike long distances—gear to prevent the rolling of ships, riveting machines, feed water check valves, steam generators, wheels for railroads and tramways, an inhaler to treat bronchitis, boot and shoe heel protectors, hair curling irons, a method for demagnetizing watches, a type of pneumatic tire, a coffee substitute, a method for extinguishing fires in theaters, and most surprising of all, new advertising methods—a rotating sign that works "even in very light airs."[13] And near the end of his life, he invented the world's first successful machine gun.

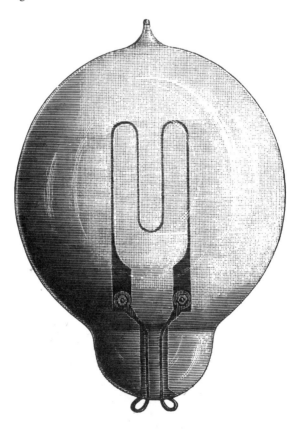

The early light bulbs developed by Hiram Maxim featured a carbonized element shaped like the letter M, perhaps to signify his last name.

Maxim's contribution to the light bulb was to improve the manufacture of filaments. Filaments, whether of bamboo or cardboard, as in Maxim's case, were converted to carbon by heating at high temperatures in the absence of oxygen until the cellulose in the material broke down, leaving a hard carbon skeleton, but uneven carbonization caused thinner sections to become much hotter when lit with an electrical current and burn out more quickly. Maxim's insight was to place a carbonized filament into a hydrocarbon atmosphere, then pass through it an electrical current that heated the filament to a bright red. The thinner and hotter parts of the filament would break down the vaporous hydrocarbon surrounding them and deposit pure carbon on the filament, building up layers of carbon on the thinner parts and resulting in a filament of uniform thickness and greater life span. As Maxim gloated, "it is absolutely impossible by mechanical means to make a carbon filament that is of uniform resistance" without his patented method, adding that Edison "had to use my process or give up the job."[14]

Maxim's attitude was prompted by the rivalry that burned between the many engineers competing in a world eager for the magic of electrical lighting, but it also shows us the problem with crediting any individual with the complete "invention" of any technology. We tend to tell the stories of inventors who, through their unique intellect and drive, produce an equally unique marvel at the climax of a story with a beginning, middle, and end. That is often how this book has told it, out of deference to individual humans' need to relate to the stories of other individual humans. But the engineering method is uninterested in this "great men" historical framework. It cares only about the accumulated knowledge, heuristics, rules of thumb, intuition, and anything else that drives problems in the direction of solutions as fast as possible, the sum of which, even for a single solution, is beyond unthinkable for a lone person to create themselves. This web of information is so vast, incomprehensibly vast, so we make it comprehensible and moving by telling the stories of individual inventors, even if this distorts the unknowable true web of invention.

Maxim is likely unrecognized as an inventor today because he lacked Edison's agile self-promotion and because, in a sense, Edison "won" and thus told the story of the light bulb's invention. But did Edison "invent" a light bulb when his company produced a brilliantly glowing but short-lived electric light? Perhaps. When we think of an invented technology, we typically imply technology that not only exists but is reproducible in a way that can fulfill the needs of those whose problem it solves. That is, it can be manufactured or mass-produced. A handful of working light bulbs in the late 1800s is a marvel, but it doesn't light the world. In this sense, the invention of the light bulb was a decades-long process of incremental changes to create a filament that can be manufactured reliably and extended beyond Edison and Maxim alone. To tell only a "great man" story hides the contributions of others who were essential to a technology's development. We can see that in the evolution of the manufacturing techniques of Maxim's light bulbs: he had on staff an artistic draftsman turned engineer whose contributions to reliable manufacturing have long been overlooked.

In 1879, just as Maxim's U.S. Electric Lighting Company was developing its bulbs, Maxim visited a machine shop to order the construction of some parts. He noticed, at a table, an African American man drafting superb drawings. The man, Lewis Latimer, later recalled that Maxim leaned over his shoulder and said "he had yet to encounter a colored person who drafted so admirably."[15] Latimer, the son of an escaped slave, learned to draft near the end of the Civil War. After his discharge from the Union Navy, he spotted an ad from a law firm that specialized in American and foreign patents, looking for "a colored boy with a taste for drawing."[16] The seventeen-year-old war veteran applied for the job and was hired at the firm. He bought secondhand books and a set of instruments with which he could teach himself the art of drawing technical models and soon replaced the departing draftsman, working late nights to coordinate with client Alexander Graham Bell after the telephone inventor had finished teaching evening classes for students who were deaf. Following more than a decade at the law firm, Latimer, now an expert draftsman, moved to the machine shop where Maxim encountered him. Impressed by Latimer's drawings, Maxim hired him as draftsman to his general assistant at the U.S. Electric Lighting Company.

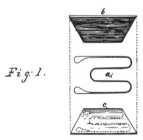

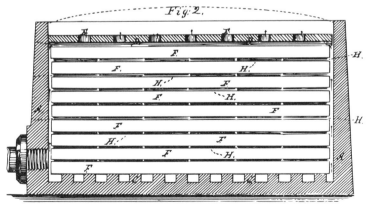

Latimer designed a method to create robust carbonized light bulb filaments.

In the small, understaffed, and underfunded company—at least when compared with Edison's bank account, filled with J. P. Morgan and Vanderbilt family money—Latimer quickly rose to superintendent of the incandescent lamp department, in charge of manufacturing the carbon filaments. He noticed that Maxim's team, despite their tidy chemical solution to the problem of uniform carbon filaments, continued to fall short of turning the manufacturing of filaments into a usefully reproducible process. When the hard, brittle filaments were carbonized, they often broke during the heating process. This wastage, of course, cost money that Maxim's lab didn't have. Latimer realized these failures occurred because the metal molds that held the paper filaments in place for the process expanded and contracted at a different rate than the paper when heated—they either pulled on the paper filament to snap the brittle carbon apart or crushed it. He scrapped these molds and instead sealed each filament in a protective envelope of cardboard thick enough that, unlike the delicate filament, it did not fully carbonize and stayed intact, but because it was made of paper, the envelope expanded and contracted at the same rate as the filament and would not tug on it. He placed these envelopes between heavy ceramic plates, covered with a top plate weighed down with sand. "The weight," he said, "exerts a continuous pressure on the layers of blanks between the plates, and thus diminishing their tendency to buckle or warp."[17]

Next, Latimer solved another pressing problem that plagued both Edison and Maxim: attaching the platinum wires to the filament. Edison used platinum-iridium clamps at the bottom of his bamboo filaments, which were cinched with a tiny screw; Maxim passed a bolt through holes at the bottom of the filament, wound a wire, then tightened a nut over it. These similar methods both destroyed many filaments in manufacturing by snapping them, or their connections were spotty. Latimer, with a colleague, dispensed with all forms "of clamps...nuts, screws, or pins, and similar accessories."[18] At the base of the filament, where the two wide sections

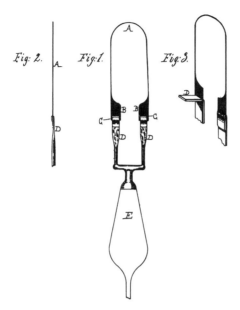

Latimer solved the tricky problem of how to attach platinum wire to a carbonized filament. This reliable connection became the industry standard until tungsten filaments replaced carbonized filaments.

at each end of the filament curved downward to connect to the platinum wires, he cut rectangular slits through which he inserted flattened copper or platinum contacts that folded over to hook the filament securely to the wire. This may seem simple today, but Latimer earned patents for both his innovations, neither of which occurred to Maxim or Edison and his team. The journal *Engineering* remarked on the "very philosophical method of constructing the carbon filaments in the first instance, and the maintaining of them afterward in a state of efficiency for so long a time."[19] As one engineer of the time reflected, "bamboo fiber was never of very much use, but the early Maxim lamps, made of paper, were very good."[20]

Latimer's methods were widely used in the industry for its first decade—a critical period that cemented the light bulb as essential—until it was replaced with "squirted cellulose," a type of artificial silk.

This new type of carbonized filament, which superseded Latimer's, was declared by Edison in 1900 to be the culmination of his twenty years of development of the incandescent lamp. It was unlikely, he thought, to ever be improved.[21] A bold statement from the PR-savvy Edison, yet Edison's successors at General Electric (GE), the final corporate form of the Edison Electric Lighting Company, knew better: Edison's, Maxim's, Latimer's, and their team's invention was not the climax of a story—it was a contribution to a process and a manifestation of a method that must continue.

Although by 1900, GE was the leading manufacturer of carbon filament light bulbs, they worried about the ceramic and tantalum filaments being developed by their competitors, filaments that used much less electricity than a carbon filament. To compete, GE intended to leapfrog the competition by using the best possible material for a filament: tungsten.

Tungsten could burn at the same brightness as carbon using only a third of the electrical power, and it could last much longer because of its abnormally high melting temperature. Yet despite this virtue, tungsten was an entirely unsuitable material for *manufacturing* filaments for one reason: it was not ductile, the property of being easily bent into new shapes, like a piece of copper, silver, or gold. Think of the twist ties that seal a plastic grocery store bread bag— the ductility of the stainless steel wire encased inside the tie allows it to be twisted hundreds of times before breaking. In contrast, tungsten, while being especially dense and hard, is brittle. A spindly wire of tungsten, when stressed to the point of bending, breaks instead. Thus it couldn't be shaped into the six or so hairpin turns necessary to construct an early twentieth-century filament. The many engineers and scientists who tried to create ductile tungsten failed, all concluding it was an impossible task.

To achieve the impossible, GE hired a young physicist, William

Coolidge. Coolidge, saddled with $4,000 in student loans from an undergraduate degree at MIT and a PhD in Germany, jumped at the $3,000-a-year salary, which doubled the compensation of his previous job, never mind the absurd task. He described tungsten as "uncompromising" and noted that "it could not be filed without detriment to the file, and was, at ordinary temperatures, very brittle."[22] Yet as he wrote to his parents, "I am fortunate now in being on the most important problems the lab has even had... If we can get the metal tungsten in such shape that it can be drawn into a wire, it means millions of dollars to the company."[23]

With a team of twenty scientists and engineers, even more assistants, and five years of work, Coolidge developed tungsten wire that was as "elastic and flexible as spun glass" and could be drawn into a piece a mile long.[24] As for how he did it, here is an abbreviated list of the steps:

1. Apply great pressure to turn tungsten powder into a fragile bar.
2. Heat it to 2,370°F, then cool with water.
3. Pass an electrical current through the tungsten while heating it to 5,790°F, then cool again.
4. Heat yet again to 2,730°F, but this time while flowing hydrogen over it.
5. Pass it through a series of dies to cold-work it, then hammer it.
6. Heat it again, then reduce the temperature gradually while drawing it into a wire with a one-millimeter diameter.

Coolidge himself described the development of these steps, which *must* be done in this order, with a telling analogy:

Imagine a man wishing to open a door locked with a combination lock and bolted on the inside. Assume that he does not know a single number of the combination and has not a chance to open the door until he finds the whole combination, and not a chance to do so even then unless the bolt on the inside is open. Also bear in mind that he cannot tell whether a single number of the combination is right until he knows the combination complete. When we started to make tungsten ductile, our situation was like that.[25]

How did he find the combination? How did he devise these steps? Without understanding the engineering method, these steps would likely seem arbitrary and Byzantine, but we can see now that it is a classic, elegant, and beautiful application of the method. Coolidge used every technique described so far. To explain his five years of research would fill a whole other book, but by looking at a few steps, we can see the engineering method in action as it pushes toward yet greater invention.

While the invention of incandescent lighting had before been driven by newly discovered knowledge of electricity, the essence of how Coolidge created ductile tungsten was based on an ancient rule of thumb: by "working" a metal, bending it back and forth repeatedly and at various temperatures, it can become ductile. This technique had been known since at least the third century but was honed to perfection in the famed Damascus blades, made of Indian wootz steel, used by *fāris* to battle the European Crusaders. For the warriors of the Middle Ages, the ideal sword blade was as thin as possible to minimize weight and material waste and to maximize sharpness while not being so thin as to break when taking the force of a swing against a shield or body. In the forging of medieval swords, as in making filaments, this made hard and brittle metal a disqualifying liability. But flexible, ductile metal, rather than being hard, was tough, more capable of bending and flexing under force to stay intact. And a series of heat treatments honed by the blacksmiths of India and the Near East turned brittle steel into pliable steel that they could forge into powerful weapons. Damascus blades were said to hold an edge sharp enough to slice a silk handkerchief floating in the air. Regardless of legends, the quality of the swords was and is renowned, but these smiths did not know why or how this process worked; they only had useful rules of thumb for working steel passed down by generations of blacksmiths.

Coolidge, though, had a clue thanks to modern science. He knew that the treatment broke up microscopic grains of crystallites. As Coolidge described it, via this process, "the crystalline structure is broken down and a fibrous structure developed."[26] He drew on long-established scientific observations, which showed that traces of impurities often destroy ductility: gold becomes brittle with as little as 0.05 percent of lead, bismuth, or tin added. Coolidge used this knowledge much like Charles Parsons used the tabulated properties of steam to eliminate less useful paths and start at a place more likely to yield success. So he began with the idea that by working pure tungsten at some temperature or temperatures, he might induce ductility.

Of course, finding that combination required much trial and error, intuition, and past knowledge. For example, the fifth step (passing it through a series of dies to cold-work the metal) came from his tour of the wire- and needle-making factories of New England. He saw how they used swaging, the gradual reduction of the thickness of a metal piece by drawing it through smaller and smaller holes; he also observed the specially shaped hammers used to shape the metal during this process. He knew that reducing the size of the crystallites in the tungsten was a key to getting ductility, and he had observed that in making ice cream, glycerin was added to stop the growth of ice crystals, so to control grain growth, he added thorium oxide to the tungsten powder, then fused it at a high temperature into a solid.

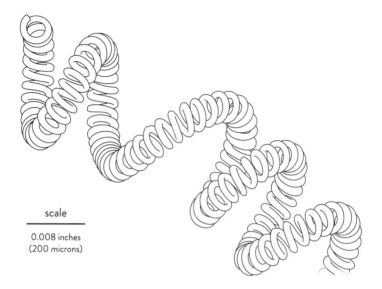

scale
_____
0.008 inches
(200 microns)

This close-up of a modern light bulb filament reveals that tungsten wire is first coiled into a helix, and then that helix is coiled into a helix. The scale bar shown is about three times the diameter of a typical human hair.

Today's tungsten filament is the most astonishing of things. It glows at some 5,000°F—likely the hottest phenomenon the average person ever observes closely (unless you are a welder). The filament starts as a wire twenty inches long and about two-thousandths of an inch in diameter. As far as the dimensions go, think of the Willis Tower shrunk to about one and a half inches tall. The filament is then wound into a coiled helix with 1,130 turns until it's about three inches long, then coiled again into a three-quarter-inch helix that you see inside a light bulb. This doubled helix packs in more wire to glow and creates internal reflections that double the intensity of the light emitted.

Since Coolidge's stunning breakthrough, the filament of an incandescent bulb has changed little, although its obsolescence approaches as tiny LEDs, light-emitting diodes, end the one-hundred-year reign of the brilliant but inefficient incandescent lamp. A white LED lamp is a purely twentieth-century device based on quantum mechanical action, about which the science of Edison's day was only beginning to speculate, harnessed for our use, yet the engineering-invention story of white LED light is identical to that of the incandescent light. Whether a nineteenth-century invention or the latest quantum-based gizmo, it is the application of the engineering method to manufacturing that puts a product in our homes and offices, and as was true of Edison and his light bulb, it is often told as a sole-inventor story.

Flip open *Big Ideas: 100 Modern Inventions That Have Transformed Our World* and you'll find a short note that Nick Holonyak invented the LED, "a durable, efficient light source that is now found everywhere, from the little red light that indicates coffee is brewing to the 22-story-high Reuters digital billboard in Times Square."[27] That's quite a leap from a single red LED to a large billboard. No mention of a chain of innovators and engineers that step by step over forty years evolved LED light into a fully manufactured invention.

White LED light can only be produced by using a blue LED that is either combined with red and green LEDs or, more efficiently, by converting part of the blue light to yellow with a phosphor. By either method, the human eye detects the light as white, but both methods require a blue LED. For decades, the only commercially available LEDs glowed red, green, and yellow, although as early as 1972, engineers had developed laboratory prototypes of working blue lights that helped to shine crisply bright and exceptionally efficient white light. A blue, and thus white, LED could not be manufactured until it was "invented" by a self-described "country boy" working not at a major electronics manufacturer but at an obscure chemical company that had never before manufactured an electronic device.[28] This country boy was Shuji Nakamura, from Tokushima, Japan. The details fill a book, but it is enough to say that much like the steps that led to Edison's first filament—the result of testing over a thousand materials—and the ductile tungsten of Coolidge, Nakamura's step-by-step approach was developed over ten years, working seven days a week, twelve hours a day, and applying the engineering method to achieve the evolution of design toward manufacturing.[29] For his work, Nakamura shared the 2014 Nobel Prize in Physics.

What harm is there in the myth of the sole inventor? Why look in detail at the evolution of a product or technology? First, the myth hides the engineering method, feeding the view that at the root of every engineering marvel is scientific breakthrough, which closes minds to the most elegant, subtle, even sublime ingenuity of engineers as they respond to constraints that arise from mass manufacturing, from the need to rapidly and reliably mass-produce a product. This ingenuity is well illustrated by a method to manufacture golf balls: the liquid center of a golf ball begins as a frozen sphere so it can be easily wrapped by miles of rubber bands. Or to create cling wrap of a uniform thickness by blowing a large bubble and then collapsing it. Or to test the reliability of a computer chip by running it at high temperature: running an integrated circuit for one thousand hours far above room temperature equals nine years of use at room temperature. The practice of engineering is shot through with these uncelebrated flashes of brilliance.

And second, not only is the method itself hidden, but people are too. To study the evolution of a product lifts from the background underappreciated people in its development—often women and people of color—and highlights the fact that engineering creativity exists in everyone. Few have heard of Lewis Latimer, yet his work improved the reliability of light bulbs for a crucial ten-year stretch. To see only a genius inventor—often a white male—as the face of engineering dissuades the next generation from seeing engineering as a creative endeavor, a profession open to all, and reduces the number of minds working to solve the dire problems our world faces. Thinking of an invention as springing from a sole inventor leads to the fallacy, often conveyed by headlines, that one climactic breakthrough conquers all problems, when every invention is only one particular culmination of countless breakthroughs, both the sensational ones and the simply arduous, over countless years.

The focus in this chapter on manufacturing highlights the fact that the line between successful and unsuccessful inventions is manufacturability. Often the bottleneck that allows the few to sweep up the fame and historical notoriety is a successful design for mass manufacturing. There are currently innumerable innovations waiting their turn to squeeze through the gap. Many innovations in energy are halted because of inadequate production methods. New materials to replace the large silicon panels that gather power from the sun with double the efficiency have been developed in the laboratory, yet panels of this new material cannot be mass manufactured—it is, says one article, "at the crossroads between commercialization and flimflammery."[30] The growing pains of a hydrogen-powered world also highlight the key role of manufacturing. How wonderful that hydrogen breaks down into benign water and oxygen, yet engineers must blast apart energy-rich methane to extract hydrogen and in the process use up more energy than the hydrogen can give them.

These possible solutions to the world's most pressing problem of energy production are stalled, for now, at the same place as Edison's filament. Many, if not most, may not have a chance to take the next step, but by understanding what it means for something to be invented, we can know that the ideas for what the future could be, with almost every detail and proof-of-concept drawn out with amazing creativity by a thousand different people, are being imagined by those applying the engineering method now. The lucky ones are those who finally get it out of the bottle.

9
CONCLUSION

What the Microwave Oven Teaches Us
about Innovation, Technology, and How the
Engineering Method Can Save the World

A search for "who invented the microwave oven" turns up a simple story—repeated in a million places, encapsulated as an entry on a timetable of science and technology, pondered in serious histories, and presented in children's books. In its most canonical form, it runs like this:

> In 1946, Percy L. Spencer, a self-educated man, walked by a radar unit—one intended for a World War II plane. A candy bar in his pocket melted, and at that moment, he devised the microwave oven as a household device to ease the workload on women, the main preparers of meals, so they could reheat leftovers or cook simple, prepackaged meals.

Punchy, vivid, and wrong.

The true story begins in 1940, when Percy Spencer inspected

what a U.S. historian called "the most valuable cargo ever brought to our shores."[1] The cargo, packed in a black metal deed box used by lawyers, departed for Halifax, Nova Scotia, from Liverpool, England, aboard the SS *Duchess of Richmond*. The unescorted "Drunken Duchess" had orders to change course every twenty minutes to avoid enemy craft, but if the ship was attacked, the box was to be thrown overboard to sink and rest forever on the floor of the Atlantic Ocean, its secrets hidden from the Nazis. Eight days after departure, the ship arrived at Halifax, where a U.S. Army armored vehicle "with submachine guns bristling from every orifice," as a passenger recalled, picked up the metal box.[2] The vehicle transported it, along with a delegation of British scientists, to Washington, DC. Soon after arrival at the Wardman Park Hotel, the British scientists unveiled their precious cargo to a curious, select gathering of U.S. scientists and engineers. It was a solid piece of copper, two inches in diameter, perhaps a half-inch thick, into which were cut circular cavities. And it was going to help the UK's Royal Air Force defeat the Luftwaffe in the Battle of Britain.

Shortly after the *Duchess* left Liverpool, a force of two hundred German bombers headed toward southeastern England, ordered by Hitler to destroy Britain's ports, stores, shipping, and oil reserves. Britain, though, had prepared for this moment over the previous five years. For most of the 1930s, British politicians and military leaders were haunted by the nightmare of a "knockout" blow from the air as aerial combat evolved from the rickety wooden biplanes of the Great War into roaring metallic war machines capable of leveling cities with withering bombing raids. In response, they built twenty-one radar stations from the Isle of Wight to the Orkneys. The Chain Home system along Great Britain's southeastern and eastern coasts cloaked the nation's underbelly with a shield of actively scanning radar waves that could detect incoming attacks by air.

Each station had four 360-foot steel transmitter towers spaced 180 feet apart that supported a net of wires—a "transmitter curtain"—that blanketed the air with radio waves.[3] Any radio waves that were reflected by aircraft approaching Britain were detected by four 240-foot wooden receiver towers arranged in a rhombus. In clear weather, each station detected aircraft up to two hundred miles away, although the average in operation was eighty miles. But even eighty miles was enough to detect German aircraft over France nearly twenty minutes before the attack made its way across the English Channel, time that the Royal Air Force could use to scramble their nimble Spitfire fighter planes. These radar reports, when combined with reports from observers at one thousand observation posts, decrypted Enigma messages, and monitored open radio communications—the German pilots had slack radio discipline and often revealed their positions on open radio—helped the British down eighty of those first two hundred German aircraft. As the battle raged for the next six weeks, the Chain Home radar system denied Germany air supremacy over Britain. But to prepare for this next wave of attacks, British scientists wanted a better picture of the skies over Vichy, France.

The Chain Home system detected only clusters of planes as a sort of blob on their screens and thus revealed no information about the number, nor could it detect planes flying below one thousand feet, but if the resolution could be increased until individual planes were detected, the balance of power in the war would change. To increase resolution, the radar needed to operate at a higher frequency: the higher the frequency, the shorter the wavelength of radio waves emitted, and the shorter the radio waves, the greater the resolution. The waves transmitted by the 30 MHz Chain Home system had almost two hundred inches between their peaks and troughs and could only detect large masses of aircraft, but that "valuable cargo," the copper disk, was a type of vacuum tube, called a cavity magnetron, that generated waves with a length of only four inches, a frequency of about 3,000 MHz, also known as microwaves. It distinguished individual planes, and equally important, it made radar portable rather than being built with towers hundreds of feet tall. Although the British had working prototypes of this magnetron, they could not mass-produce them, because the Nazis had cut the UK off from the Continent. The British needed U.S. technical expertise and production capability.

The first U.S. technologists who viewed the magnetron could not help. As they explained, they were research scientists, not manufacturers, but one of them suggested they visit "a little tube plant" near Boston.[4] "They've got some people out there who are pretty knowledgeable." The plant was owned by Raytheon, a small manufacturer of radio and vacuum tubes, which often struggled to turn a profit. The "pretty knowledgeable" person was their resident genius, engineer Percy Spencer. So clever was Spencer in making vacuum tubes that he boasted "given a milk bottle, a tin can, some bailing wire and a bucket of whitewash, he could make any kind of an electronic tube."[5] And indeed for nearly twenty years, he helped Raytheon scrape by with his novel tube designs. He taught himself electrical engineering while serving in the navy: "I got hold of textbooks and taught myself while I was standing watch at night."[6] At Raytheon, he continued this self-education, assigning one of his engineers to gather, every week, all new vacuum tube patents. This engineer then reviewed each patent by writing a paragraph-long critique with an emphasis on how it applied to Raytheon products. Spencer studied these reports with an eye to enhancing the Raytheon product line.

When the British delegation brought Spencer the magnetron for advice on manufacturing it, he asked to borrow the device over the weekend. "If you want me to give you any help," Spencer told them, "I need to take this thing and look at it and figure out what could be done."[7] The British at first refused. It was top secret, they told him, and its possession would endanger him. It could well be that the delegation had been followed and Spencer might be assassinated if he made himself such a high-profile target. But with reluctance, the British scientists relented for the sake of gaining their new defensive tool as quickly as possible. Once alone with the magnetron, Spencer turned the device over in his hand and studied it. To him, the design was "awkward" and "not practical."[8] It would be, he thought, impossible to manufacture in any volume, requiring "much equipment and so many skilled people." The cavities carved into the block of copper had a tolerance of less than one ten-thousandth of an inch; the slightest deviation in the cavity's size would change the frequency of the radiation. Given the parameters, a master mechanic would need a week to finish a single magnetron. Spencer's most limited resource was time, and the circumstantially emergent notion of "best" was demanding. But he had an ever more thorough scientific understanding of electromagnetic radiation specifying his goals and entire industries full of rules of thumb that fueled the development of mass manufacturing in a hyperindustrialized world. As it was, a successful magnetron, he thought, was "at the mercy of human skill," but if he could remove that element, perhaps the magnetron could be mass-produced into an arsenal for the Royal Air Force.

When the British returned on Monday, they found a smiling and optimistic Spencer. Instead of a solid block of copper with precise cavities cut by a master mechanic, he proposed assembling the body out of ten or so thin sheets of copper, about a sixteenth of an inch thick. In doing this, he followed a time-honored rule of thumb: break complex problems into smaller, more manageable pieces. Here, he did this literally. A part of the cavity was punched in each of these sheets, so that when the sheets were stacked on top of each other to construct the magnetron, they created a cavity throughout. These sheets could be rapidly punched with large presses operated by semiskilled workers, precisely cutting the sheets within the required tolerance using pretooled dies. "Boom, boom, boom," as a Raytheon engineer described it.[9] Spencer had taken all the precision off the hands of the humans crafting the magnetrons and given it to the die, which could be used thousands of times. To hold these stacks of copper layers together, he inserted between each layer a wafer-thin sheet of silver. When passed through a furnace, the silver melted and fused the copper sheets into a solid block of copper, much like the prototype brought across the ocean. With this method, Percy Spencer remade the cavity magnetrons from an artisan-crafted instrument to a mass-produced device that could help win a war.

As the war progressed, Raytheon manufactured one hundred magnetrons a day, then one thousand a day, and then two thousand five hundred a day at its peak. Of the one million magnetrons used by the U.S. Navy, Raytheon produced half and some 80 percent of those used by the U.S. armed forces as a whole. So many that the production manager kept track by listing monthly production by the pounds, rather than number, of magnetrons manufactured. The greatest impact of Raytheon's magnetrons was in their shipboard SG (Search Gear) Radar systems in the Pacific theater of the war.

During the war, Raytheon grew from a small company with $3 million in annual sales to, by the last year of the war, a behemoth with $180 million in sales—more than $800 million over the course of the war, about $13 billion in today's dollars. But as the war ended, Raytheon's executives knew that their lucrative military contracts would vanish, causing the company's revenue to drop by at least 50 percent. If Raytheon was going to keep its newfound dominance, it needed to come up with something it could sell to the new consumer marketplace of the Pax Americana.

It isn't clear how or when it dawned on Spencer that a microwave-emitting magnetron could generate heat or even that he was the first to think of it. For sure, there was no candy bar moment—a detail invented by a *Reader's Digest* writer in the 1950s—but during the war, it was common in winter for Raytheon engineers to walk past banks of magnetrons operating in the open air and warm their hands on the heat they emitted.[10] Spencer took note of the "aging racks" where magnetrons were hung up to run continuously, sealing the vacuums and increasing their reliability, and envisioned hundreds of uses for microwave heating. "Goodness knows," he told *Fortune* in 1946, "eventually the field will be tremendous. We can't begin to think of all the uses—ink drying, soda straws, tobacco curing... There's so much to be done it drives you frantic."[11]

We can now begin to see how that punchy, vivid, and wrong origin story of a candy bar melting oversimplifies the story. It turns it into a linear path from inspiration to invention.

Percy Spencer + radar + melted candy bar ⟶ Microwave oven in every home

The radar origins make clear that this mythical story obscures the engineering method. At first glance, it hides the insight needed to take advantage of a chance observation and the diligence needed to develop it. It overshadows the extensive and detailed research that led to many iterations of an eventually ubiquitous invention. As one would suspect, compact and convenient household microwave ovens, technically "consumer" ovens, were preceded by large industrial, or "commercial," ovens created by years of engineering sparked by the possibilities Spencer saw in that copper box. Our simple, linear diagram can now become a ladder: the bottom rung being the first observation of microwave radiation's unique heating properties, then an inevitable climb up this ladder of progress through the first clunky prototypes, then, at the top of the ladder, the essential and streamlined household ovens of today.

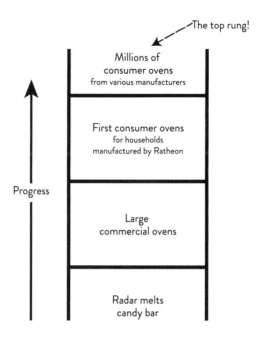

This image, albeit slightly more sophisticated, still distorts history in such a way as to do us a disservice. It suggests that engineering follows a linear trajectory, and it leads to the belief that once an engineer has "launched" on this trajectory, they will inevitably arrive at the only possible solution to the problem at hand, at the style and operation of microwave oven that all follow today regardless of the bells and whistles on the outside. It suggests that the microwave oven's invention and final design were *solely* a technical matter of applying the observation that radar generates heat. The fuller story of the oven's invention dispels the myths.

Reflect for a moment on Spencer's comment that "there's so much to be done," written before the first microwave oven existed.

Of course, to an extent, he was talking up the technology as a matter of marketing and publicity with a popular magazine, but his goal as an engineer wasn't to create an oven. It was to develop a new heat source that had many uses, even to the point of displacing fire. This might lead us to modify our diagram of its development to picture it not as a ladder but as a tree:

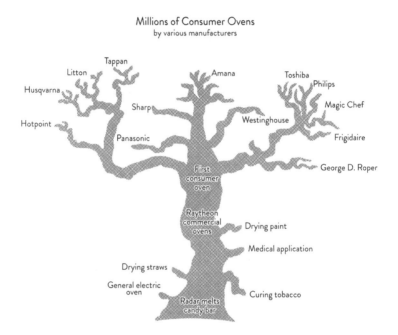

The tree's root is that radar melting a candy bar. Near the bottom, tiny branches that never grew much—drying paint, medical therapy, industrial applications—and then, in the solid trunk of the tree, Raytheon's commercial oven, its first consumer oven, and the growth of the thousands of consumer models, all operating in the same way but with different chassis or external controls. Yet this image, although better than the ladder, still distorts the idea of how technology evolves: it implies that while there were many ideas *at the beginning*, these died out quickly, and that the oven, once conceived, moved forward—a type of determinism. We think of the microwave oven story with a kind of retroactive teleology, imagining that all along, the device was meant to be used in the way we use it today. Yet the consumer oven was never the intended outcome: the goal was a large, commercial oven that would streamline the production of food in restaurants. The modern microwave oven is a *failed* version of what the Raytheon engineers were trying to build.

Following the first steps of the engineering method, Spencer and the Raytheon staff identified the problems to be solved. In the almost fully electrified developed world, people had gotten used to the instant gratification of light summoned with a switch, travel measured in dozens of miles per hour, food and drinks kept frozen and chilled in home refrigerators, antibiotics eliminating the threat of bacterial diseases. In the postwar boom, there would be an almost exasperated expectation that all aspects of modern life be made faster and easier, that there must be some new, better way to do everything. The fact that curing materials, drying liquids, and cooking food still had to be done with the primordial heat sources of fire and sunlight had become a problem to be solved. As always with an engineering solution, the notion of best was fluid: it was changing with time.

The science, as it was, had been laid out before them. There's nothing new about heating with electromagnetic radiation, which encompasses all light and heat. Microwave heating is only a branch of a much larger tree whose root is the beginning of technology—using fire to cook food. Visible and infrared radiation have always been used for cooking, from the glowing coals of ancient times to today's electric broilers, heat lamps, and Easy-Bake ovens. Microwaves are unique in that these original types of radiation cannot penetrate the surface of most foods, so they work only on the outside, with heat diffusing from there to the interior. But microwaves, being a type of radio wave, can penetrate the outer layer of food, in the same way that they pass through the walls of a house or a car to reach a radio device, and heat the interior directly. They do this by setting the polarized molecules of substances like water, oils, and other food components into rapid motion. Since a molecule in the middle of a piece of food can receive this energy as readily as one on the exterior, microwaves are sometimes said to cook food from the inside out. In practice, they are generally absorbed in the outer inch or so of a piece of food, which is why reheating a whole deep-dish casserole in the microwave can leave it cold on the inside. But in the case of Spencer's microwave heating, no science was necessary to start the ball rolling. By observing its application to the problem of radar engineers' cold hands in the wartime winter, the fact that it was a useful source of heat was enough to brainstorm solutions. The power of science would have its applications later.

But in suggesting that Spencer and Raytheon are the protagonists of the story of the microwave oven, the tree diagram imposes its second distortion: the Raytheon engineers weren't isolated geniuses exploiting a discovery no one else thought of. The convenient physics of microwaves were no secret after the industrial-scale military use of radar, and the notion of using microwaves for heating was, so to speak, in the air. GE, Westinghouse, and RCA were developing microwave devices for processing plastics and rubber, textiles, and wood, and GE was working on its own version of a microwave oven. Electronics-industry trade journal articles described the prospect of "sterilization, blanching, and cooking of food stuffs" by microwave radiation.[12]

Spencer and the Raytheon staff were initially focused on heating food and thus started by building prototype ovens whose design and power would make modern bystanders uneasy at best. One began as a garbage bin lying on its side with a radar tube sticking through a hole in its back, which they used to pop corn and experiment with eggs, which would explode when heated rapidly. The most important insight from their experiments, though, was that whatever they cooked was always finished in a hurry—much faster than with conventional methods.

Making a device for sale to the general public was something new for Raytheon; most of its equipment was built to be run by trained specialists, like military officers, who could be trusted to read the manual and handle an instrument carefully lest they find themselves on the wrong end of a dangerous and expensive piece of equipment. But the oven would have to be much more foolproof than any military device if it was going to be put to work for the everyday citizens Spencer imagined making use of the oven in industrial food processing and restaurants. Operation had to be as automatic as possible, for many users would be short-order cooks who would likely experiment with new ways to prepare a dish. Their essential problem in designing the oven was finding—as described in an article at the time—"a suitable frequency, cavity size, food stuffs, and proper orientation of the food" to get uniform heating.[13]

Raytheon was up against the classic dilemma of any design engineer: choosing trade-offs among incompatible constraints. They had to balance requirements that are only partly compatible, turning the dials on each parameter in order to form the best solution to the problem at hand, made even harder by the fact that there wasn't yet a cultural consensus on what a quality microwave oven could do for its user. The main difficulty, still suffered by all home microwave cooks, was the frequency of the radiation. The magnetron that they started with produced radiation with a wavelength of about twelve centimeters, or five inches. This was convenient, because, as Spencer noted in his 1945 patent application, microwave heating works best when the wavelength is comparable to the size of the food being cooked.[14] But that meant the wavelength was also comparable to the size of the oven cavity, which in turn meant the microwaves could easily set up a standing wave inside: the microwaves reflected from the sides of the oven canceled and reinforced each other to create an unvarying wave, which causes uneven heating. As one engineer said, "If you put a hot dog in that, it would be cooked in the middle and raw on the ends."[15]

Why not just change the frequency of the radiation or make the oven a different size? Outside cultural forces stepped in to impose their own restrictions and define the metrics for the "best" oven: it had to be compact enough to fit in a restaurant kitchen. This meant a cubic oven capacity no bigger than about twelve inches on a side. As for the wavelength, even before World War II ended, microwave devices were being developed for such uses as FM radio, long-distance telephone relays, aircraft guidance systems, and even facsimile transmission. The Federal Communications Commission realized that it needed to allocate bands of wavelengths for specific uses to minimize interference, and the 915 and 2450 MHz frequencies were set aside for "industrial, scientific, and medical" purposes, including microwave heating.[16] This gave the microwave oven engineers a choice between either a 12.9-inch or 4.8-inch wave, respectively, which left 2450 MHz as the only viable option for a countertop oven since it would create more frequently spaced standing waves than the 915 MHz wave, which would reach across the entire oven cavity. The best solution, then, was entangled with shared industrial needs, turning the design decision away from a purely technical choice, as simple models and stories imply, and making it into a complexly human one that was not only best for microwave heating but for the creation of a new world of electronic communication.

To tackle the design of their oven, Raytheon engineer Marvin Bock needed to mitigate, if not eliminate, the standing waves that caused uneven heating and disqualified the microwave oven as a replacement for the gas or electric range oven.[17] To do this, he needed to disrupt the standing wave. The most promising solution was to vary the size of the oven cavity while the food was being heated. First, he tried using a small motor to move one wall of the oven back and forth, but this proved to be too complex and expensive for mass manufacturing, partly because of the motor and partly because of the radiation that leaked out of the joint between the moving wall and the body of the oven. Bock realized that he could change the oven size virtually: from the upper wall of the oven, he hung rotating rods turned by a motor. As the radiation bounced from top to bottom of the oven, this "mode stirrer" alternately blocked and allowed certain parts of the waves, creating a constantly shifting set of standing waves. While the solution was not perfect—even today, no microwave oven warms food completely evenly—it distributed the microwave energy through the cavity well enough to have a shot at Spencer's ambitions.

Now, with all these elements in place—the tube, its power supply, and the mode stirrer—Bock turned to the real problem: cooking food. First, he tried popcorn. "Refreshing corn popped takes twenty seconds is good," with 80 percent of the kernels popped, he recorded. Next, he looked at vegetables. For potatoes, he noted that "the flavor was good but the potato was not crisp. The time required was one minute." Other notebook entries: "Brussels Sprouts, 1 minute 15 seconds, no water was used. The flavor was dry and not good. Brussels Sprouts frozen, the flavor was good… Mashed frozen potatoes, the taste was good but they did not brown. Time required: 1 minute." Then came the ultimate test in meat. "Noted that chicken Fricassee took a minute and tasted good; steak doesn't brown and a thin steak works better for browning. Thick steak tastes good, retained juices and flavor." Here we see how engineers in the decade after World War II envisioned the use of a microwave oven: to replace the conventional oven, to cook, as the images from early patents show, whole chickens, lobsters, and roasts. And most importantly, it was *not* intended for home use.

The first production microwave oven weighed in at 670 pounds, stood sixty-two inches tall, and measured nearly two feet deep and wide. To install it, an electrician had to put in a 220-volt line, and most of its volume was taken up by the water cooling system, which required a plumber to connect the unit to the kitchen's water supply. The remaining space contained the roughly one-cubic-foot cavity for cooking. This first oven sold shortly after the war ended for more than $2,000, the equivalent of about $25,000 today.

The driving force for an oven of this size and power was not a technical limitation but a choice that reflected the wartime mentality of Raytheon's engineers. Throughout the war, they had worked on methods to manufacture magnetrons as fast as possible. Every bit of time they lost was one less radar manufactured and might well mean one more American soldier killed. They read their own concerns and motivations into a restaurant's use of a microwave oven, with their vision of the oven as a tool to reduce the time restaurants spent preparing meals. So in keeping with Raytheon's military background, when they designed the first commercial microwave ovens, they built the biggest, fastest, and baddest oven possible.

To demonstrate the oven to restaurant owners, Raytheon sales staff would zap a cob of popcorn and wave it around moments later, fully popped. Other demonstrations included cooking a six-pound roast in two minutes, a hamburger in twenty-five seconds, and a flat slab of gingerbread batter into a springy dome in twenty-nine seconds. Their pitch to a restaurateur was the oven's ability to minimize wasted time and material. If they were short a few ribs when serving a large group, the kitchen staff could quickly roast a few in the microwave oven. Chefs could even develop special microwave dishes that were cooked on demand in serving dishes. Yet while the feats of speed cooking were impressive, few at the company stopped to consider whether those extra minutes were worth the massive investment they made necessary for a local restaurant that wanted to join the future of cooking, as they saw it.

Usually, at this point in the conventional tale of the microwave oven's inevitable rise, the story jumps to the successful consumer oven, yet for a moment, microwave heating was not for cooking food but warming human flesh—an interest of the newly appointed Raytheon president, Charles Francis Adams Jr. Underneath Adams's quiet, bean-counting demeanor lay the soul of a true technophile. When Raytheon brought him on board in 1948, he vowed to avoid depending exclusively on military contracts, so he pushed non-military, commercial applications of microwave heating, which his eccentric interests directed toward medical applications. He loved to ask Spencer, whom he greatly admired, to build microwave devices to warm aching muscles; he would quietly take them home and test. Under his regime, Raytheon brought all the way to market a microwave muscular diathermy unit that promised "penetrating energy for deep heating" that created in the human body a "desirable temperature ratio between fat and vascular tissue." Their greatest worry was not irradiating people with microwave energy but, as an ad noted, "No tangle with television!" They assured users the unit used "frequencies way, way above the television range. No danger of interference."[18]

It's easy to laugh at the thought of turning our kitchen microwaves into muscle-relaxing tanning beds and to place this commercial failure on our ladder or tree diagram as perhaps a broken rung or gnarled branch. Yet this failure highlights the fact that the story of the consumer microwave oven is not a simple, unimpeded, inevitable rise. In fact, it was only after this failure that Adams turned toward developing a home appliance after a false start with restaurant clients. The intent was to solve the problems of American house-wives, expected to prepare the family's meals. Raytheon intended to ease the workload on women, the main preparers of meals, so they could reheat leftovers or cook simple, prepackaged meals. To that end, Adams continued the development of the microwave oven, hiring an industrial-design firm to create a new microwave oven that would be attractive to housewives. The designers gave it an elegant overall look and components such as knobs shaped to fit the average metrics of a woman's hand. But if Raytheon was going to reimagine the microwave oven as a household kitchen appliance, they would need to do what often brings out the greatest creativity in engineering: making it on the cheap. To engineers accustomed to designing equipment for the U.S. military, where reliability was paramount and price was not an object, such an approach was alien. To bring in a fresh perspective, Raytheon used their wartime wealth to buy high-end refrigerator manufacturer Amana. Like Marvin Bock on the original model, the combined teams began with the magnetron tube. The Raytheon tubes cost $300, which alone was over half the $500 price tag that Amana's founder and president, George Foerstner, wanted to put on the oven. The impact on the initial cost was enough to prohibit using this tube, but even more damning was that if a warranty repair occurred, all profit would be lost. The answer was in the global economy emerging from the ashes of World War II, from the nation across the Pacific that Raytheon had worked to defeat only years earlier.

The New Japanese Radio Company (NJRC) designed a tube for cooking that was just good enough, with their notion of best defined by the needs of a recently industrialized war-ravaged country and economy, for the operation of a home kitchen oven that cost less than $25. The Raytheon tube had thirteen separate metal parts, including ten cooling fins, which had to be carefully put together.

Echoing Spencer's solution for magnetron production, the NJRC-designed tube was punched from a single slug of metal in a die—cavities and cooling fans formed in one swoop. The cavities were not as precise as those of the Raytheon magnetron, but this was not for a radar system that needed precision to resolve individual aircraft in the sky—it only needed to meet the tempered ambitions of the home microwave oven. The Raytheon tube was capped by a glass envelope, which could melt; the NJRC tube used a durable ceramic cap. The Raytheon tube used an expensive Alnico magnet; the NJRC engineers used a cheap ceramic magnet whose performance dipped as its temperature increased so that after the first minute or two of operation, the magnetic field dropped and the tube's power output to the oven decreased. Again, this wouldn't do for Raytheon's military-grade radar but was good enough for cooking a hot dog. Finally, the NJRC tube brought a critical benefit to the oven. A lead engineer at Raytheon recalled that it had "a very modest heater power, 65 watts... And it was about 65 percent efficient, which made it fit into a 15-amp household circuit."[19] No special wiring had to be installed; a consumer could bring the oven home, plug it in, and use it right away. The reduced power meant, of course, that the oven wouldn't be able to cook nearly as fast as Raytheon's original model, which could blast a beef side into a well-done steak in minutes, but the younger Foerstner had a clear sense of how fast was fast enough.

In 1970, Raytheon and other U.S. manufacturers sold forty thousand microwave ovens at $300 to $400 apiece, and by 1971, Japanese manufacturers had begun exporting low-cost models priced $100 to $200 less. Sales increased rapidly over the next fifteen years, rising to a million by 1975 and ten million by 1985, nearly all of them Japanese. In the decades since 1970, the microwave oven has followed the familiar path from high-priced wonder to cheap, ubiquitous necessity.

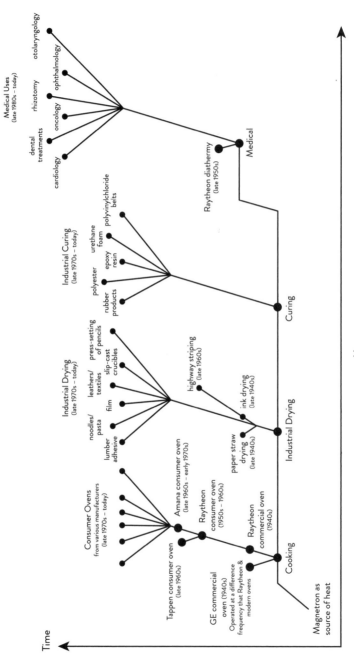

This history shows us the problems with visualizing the development of the microwave as a ladder or even as a tree. A tidy story prunes too quickly the parallel developments and squeezes the story into one of success created by reading backward from the present—what a historian would call a "Whig" history. These diagrams, which represent the simple origin myth of the microwave oven (a candy bar melts in Percy Spencer's pocket and voilà! he invents the oven), hide the rich details of how engineers work. While the story is often a necessity, the actual invention of the microwave oven was not a story; it was a system, a process. Being such, a much better way to visualize the development of the oven is via a diagram using cladistics, a method used by biologists to chart evolutionary changes and relationships. It captures complex, conditional progression, always recalculating and rearranging for both creative and practical purposes, without the bias of inevitability.[20]

Graphed on the vertical axis is time and, on the horizontal axis, applications of microwave heating, which branch as a new application evolves from a previous use. This diagram shows many (but not all!) of the products and processes that have emerged from heating by a magnetron. If this diagram confuses, that's part of the intent: a reminder of how we simplify the evolution of a technology. On this diagram, you see the first applications emerge *simultaneously*. The GE microwave oven, which operates at a different frequency, appears close to Raytheon's first oven. A little later, Raytheon's diathermy appears, contemporaneous with the second version of the oven. And the diagram highlights how many of these initial uses evolved into current industrial processes, none of which are mentioned when we talk of the microwave oven despite being essentially the same technology based on shared rules of thumb, intuition, and applied solutions. To create the ladder or tree diagram, we select only a subset, which distorts our thinking of the oven's development and more generally of how technology evolves. Why, though, tell the true story of the microwave oven at length? Why reveal, even revel in, the dead ends and failures of developing the oven and chart its development with all kinds of microwave heating? What is the harm with the simple story? After all, it's just a kitchen device.

Simple ideas of technological development foster myths that disempower us as creators, users, regulators, and learners of the engineering method. When we prune from the story all failures, false starts, and inconsistencies, we demote them on a tree diagram to a small branch or remove them altogether in a ladder diagram. But this strips away all the richness that reveals how the engineering enterprise works. And these simple stories of technology privilege science and thus distort how public moneys should be spent: we direct science toward applications, when we should spend on pure science that generates the powerful rules of thumb used by engineers.

Hidden, then, by the simple story are many of the social and political changes of our era. The full story of the development of the microwave oven encapsulates much of the story of technology in the second half of the twentieth century. It reveals the great influence of World War II technology in our lives, charts the rise of electronics as the greatest force in technological change, reflects the growing sophistication of advertising and marketing, and, as noted, exemplifies the great change in women's roles due to the industrialization of the household.

The simple story of the oven misrepresents the nature of the problem and whose problem Raytheon would be solving. They intended to ease the workload on women, since they were the main preparers of meals. Raytheon's designers and engineers saw themselves as part of a booming industry doing a service to America's overworked mothers. And of course, there can be no dispute that at the beginning of the twentieth century, housework generally became much less backbreaking. But for whom did appliances make work easier? And for whom did they save time?

To probe this, compare a meal prepared today to a meal prepared in the nineteenth century. In that century, most families ate stew, just a big pot of meat and vegetables cooked in liquid for a long time.[21] The process of making the stew took labor provided by both sexes, each determined according to traditional gender roles but comparably demanding. A man used handmade knives to butcher an animal; a woman carried water to the house in wooden buckets held together by leather likely tanned by her husband. She cooked the stew, made of vegetables from her garden, over a fire using wood chopped by her husband. She thickened the stew with grain husked and threshed by her husband. Any scraps or garbage that were not used were moved outside—likely again by her husband. Now, in the era of the microwave oven, we buy food from the grocery store, throw it in a manufactured steel pan, flick on a burner, cook dinner, and toss the scraps into a garbage disposal. Note what happened to housework: technology liberated men from their traditional roles while leaving women with their responsibilities, and the expectations of cleanliness—now the sole duty of women—were often raised by the power and ease of the household technology available to them. In the past, a children's chore was to beat the rugs outside a few times a year; today a woman typically drags a vacuum cleaner across the carpet on a regular basis with the children moved out of the way.

Of course, none of us would want to do away with our vacuum cleaners or kitchen microwaves. Modern cultural expectations demand that we maintain levels of productivity and cleanliness that would be impossible without them. But gender norms that persisted through the postwar boom's breakthroughs in household technologies skewed responsibility for home upkeep toward women in a way that did them few favors, despite the designers of those postwar appliances intending the opposite. They unknowingly fell into the same traps that were laid for the early inventors of bicycles and color photography by their limited perspective of the problem they were solving. Predictably, holistic and cross-discipline understandings of complex problems are what bring engineers closer to the starting line of the marathon they're trying to run.

When we think of technological progress as inevitable—that the products we have that make the world what it is today were pre-determined by the genius of inventors—we overvalue the end products of engineering to the exclusion of the process that this entire book is about. Such thinking absolves engineers of any failures and turns the story of the oven into a cultural staple, sidestepping the much more useful and interesting fact that it was a wildly successful failure. This myth of inevitable progress through inspired genius hides the fact that the reality of engineering is radical pragmatism: what matters is that someone managed to solve enough problems to create a new device or system. The messy and complicated engineering method is, by necessity and nature, radically pragmatic and creative, able to constantly adapt to a changing world.

To believe in the myth of inevitable progress through inspired genius supports even more expansive myths that technological advances will abruptly, immediately, and dramatically reshape the world, that a technological fix exists that solves all ills. Technological innovation certainly changes the weight of the various factors and alters costs and benefits, but the fix alone cannot solve a complex problem. There must also be buy-in by the public—the solution must fit the mores of a society. The case of nuclear power alone proves this point. In the 1950s, nuclear power was the ultimate fuel source; in a utopian vision, it would, in one fell swoop, eliminate dirty, expensive fossil fuels. Yet today, it supplies only 10 percent of global electricity. This is partly because the electricity is more expensive than conventional sources due to the high cost of constructing a nuclear plant and partly because the public never fully adjusted to the risks associated with nuclear energy.

These simple pictures also diminish public belief and support for engineering. The ladder or tree image reinforces the common but faulty belief that engineers start with an idea or goal and work toward that specific goal only after the spontaneous moment occurs. This teleological fallacy imagines that the development of a technology or an object follows a straight line, that an engineer starts with an idea unique to their genius then heads toward it like a bullet, which creates an unreasonable expectation in the public of what public or private investment is paying engineers to do and risks a loss of faith in the profession when we need it the most. Will we ever believe in the safety of a particular model of plane after a tragic crash or nuclear power after the Fukushima plant meltdown? Knowing the engineering method reveals why we might trust these technologies again. As engineers create, they acquire more and more information, rules of thumb, and intuition, and they see the value society attaches to a particular outcome—the problem we want or need solved—change.

We have a responsibility to understand that there is and never was anything inevitable about engineering. There is no "must." Which means that we have the responsibility to control it. Simple stories create an image of technology shaping our lives with a ruthless logic all its own, a great whirlwind of innovation that sweeps through our world, granting blessings and sowing havoc as we and technology blindly interact with each other. This is only half true and, because of this, dangerous. Its truth lies in the degree to which technological achievements *do* affect our lives. Never before has such a complex web of technology permeated a culture. For sure, in every century, some marvel reshaped the world—the printing press, gunpowder, the cotton gin—but only in the twentieth century did these wonders unite into a comprehensive system that overtakes us. Technology aids us in our work, assists us in the tasks of daily life, and, when we are tired, entertains us in ways that are impossible to escape and, to an extent, force us to adapt to them. But when reactive adaptation is our only end, it removes from public discourse a vital discussion about technological systems—in the words of engineer turned historian Thomas Hughes, "to exercise the civic responsibility of controlling those forces that in turn shape our lives so intimately, deeply, and lastingly."[22]

The first step to controlling those forces is to replace simple, mythical images of technology with a rich, nuanced understanding of how engineers create with the engineering method. One computer scientist advocated the maxim "Program or Be Programmed"—learn to control a computer or it will control you.[23] The same applies to simpler technologies: knowing the fuel used to generate your electricity, the source of your water supply, or what happens to your recycling empowers you to advocate for change because you know that the solutions being used on your behalf were chosen for complex but identifiable reasons, and when you reexamine the variables applied to the engineering method, new solutions are always an option.

A first and necessary step in taking back control is to understand how a technology was developed, how self-driving cars can become lethal, or how gene modification has technical limits. This understanding, although necessary, is only the first step. The engineering method is a tool, albeit, as demonstrated throughout this book, an extraordinarily powerful, often revolutionary one, but it has no morality, no sense of human consequences, no desire to seek what is best for humanity. As a pioneering historian of technology wrote in 1934, "Railroads may be quicker than canal boats, and a gas lamp may be brighter than a candle, but it is only in terms of human purpose and in relation to a human and social scheme of values that speed or brightness have any meaning."[24] This quote highlights the fact that doing engineering is an ethical activity, that moral choices happen at every step.

Although I have presented utilitarian arguments for any moral questions encountered—advocating for inclusion by noting that all minds are needed rather than using a deep, moral argument—this book lays the essential foundation for the reader to ponder these deeper questions. This fine-grained description of the engineering method sweeps aside the two most common (and useless) extremes used to frame ethical and moral questions of technology. In one extreme, value questions enter too early. Philosopher Hannah Arendt argues that the engineer—and by implication technology—is not a master of their fate.[25] The engineer, to Arendt, simply opens Pandora's box, which unleashes horrors that threaten the human race. The other extreme glibly sweeps aside ethical and moral considerations. Techno utopians—they mostly live in Silicon Valley but have been with us since at least the nineteenth century—proclaim that technology will solve every woe.[26] We see, with a firm grasp of the engineering method, that these are caricatures that remove the human aspect of being an engineer—the struggle to create under the most adverse of conditions, by using what economist Julian Simon called the "ultimate resource," human ingenuity.[27] To build an ethical and moral framework, we need first to understand what an engineer does and how they do it and then turn to the larger, deeper questions to figure out how to make the world a better place.

AFTERWORD

My childhood vacations more than fifty years ago planted the seed that grew into this book. Wherever we visited, my parents, a botanist and a theater professor, organized factory tours. My brother, sister, and I watched spellbound as a Canadian factory sliced tree trunks into lumber or a plant in California poured corn into thousands of cans a minute, and we held our noses in the putrid air of a Kellogg's cereal factory in Michigan. While both parents wholeheartedly approved of these tours, they were of special interest to my father. These miracles of manufacturing fascinated him, although he often exasperated my scientist mother with his nearly complete lack of scientific knowledge. I still recall the disappointment of my father when he learned we were too young to tour an automotive assembly line.

It was natural, then, for me to enroll in engineering school, propelled partly by this exposure to manufacturing but also because my older brother, while home from college at fall break, groused about the small salaries offered freshly minted biologists, his major at the time, in contrast to the larger starting salaries of engineers—a strong appeal to my cash-poor eighteen-year-old self. In the fall of 1980, I arrived on campus at a small engineering school in northern Michigan; it astonishes me now to realize that I applied to no other college, nor even considered another major. From the first moment, I was immersed in chemistry, physics, and mathematics classes; at that time, there were no engineering classes in the first year. To the great surprise of my mother, I thrived: a four-point average in my first quarter! To recall this is not to cast aspersions on my mother. She and my father were never less than supporting and encouraging, because no one, not even me, would bet on me excelling. In high school, I displayed, at best, a solid adequacy, not really even a spark of promise, certainly no intellectual fireworks. Yet in that first year, I shone as a dean's list extraordinaire. Then, in the second year, I entered my first engineering class.

I wish I could report that this class grew that seed planted by those family vacations, that like a skilled, clement gardener, the class nurtured my nascent interest in engineering until it gradually bloomed into a deep, rich understanding of engineering, but instead, the class flummoxed me. The problems were vague, poorly worded, imprecise—how I longed for the clarity of the first-year science courses. The class used science in an odd way: instead of drilling down to the fundamentals, they grouped things together and thought in what I then called "pseudoscience," while I wanted to start with atoms and reason from there! Even the grading baffled me. When they returned the first test—and still today, forty years later, I recall the pale color of the blue book in which I wrote the exam, and I can feel its soft, cheap paper—I saw scrawled on the cover a giant, red, circled twenty-three. My friend, sitting next to me, showed me his: it was marked with a fifty-eight. His scores on our chemistry, physics, and mathematics exams always trailed mine, often by quite a bit. Engineers, I thought, even grade oddly, using an inverse scale. Alas, engineering professors graded like all professors: we had both flunked the test, me most profoundly.

The astute reader can see in my puzzlement the elements of the engineering method described in this book, its essential aspect seared into me by a first failed encounter with engineering: that the engineering method is not the scientific method, that engineers work without complete knowledge, and that they use sometimes-wrong rules of thumb. I dropped the course, thought about my future, and reenrolled the next semester. I earned an engineering degree—and then two more—and have been a professor teaching engineering for over thirty years.

This launched me on a lifetime journey of marveling at technological handiwork. At first, I admired the spectacular engineering achievements I encountered in childhood—the dazzling assembly lines, the majestic rocket launches of the Apollo space program, the tremendous skyscrapers of New York as my theater professor father dragged us from show to show—but later I admired the quotidian, the disposable, the ephemeral (a can, a plastic bottle, a Q-tip). Still to this day, I am awed by our engineered world. I light a burner on my oven and wonder where *exactly* the gas comes from, flip a switch and ponder the source of electricity. Ketchup packets in a fast-food restaurant intrigue me—how do they seal them without wastage?

While these products of engineers were apparent to all, I noticed that the *practice* of engineers was sidelined in our great scientific institutions and buried by myth in popular culture. Our research universities operate using a model first propounded by Vannevar Bush in his 1945 manifesto, *Science—The Endless Frontier*.[1] In it, Bush extracted a lesson of the technological achievements that won World War II—the atomic bomb, radar, the mass production of penicillin[2]—science first, then application, an "applied science" model often called the "linear model." At university, this privileges scientific inquiry, then hides the role of engineering after scientific discovery. A version of that linear model informs popular accounts that feature a lone inventor—John Harrison and his longitude-measuring watch, Edison and his light bulb, or Philo Farnsworth and television. These popular stories hide details of engineering behind the romantic story of battle of that "lone" inventor to be recognized, the drama to overcome the odds.[3] This constant drum-beat of the linear applied model beloved by academics and popular accounts shapes the thought even of those *trained* in science. I recall at a dinner party, a physicist—a physicist!—asked what exactly engineers do. I'd describe his notion of their work as, at best, fuzzy, while his attitude questioned their need in our modern age, where scientific knowledge advances rapidly.

While that linear model clouds his vision, another reason that engineers are hidden is their success. The hallmark of good engineering is invisibility—we rarely think of our furnace, a jet's engine, or the purity of a pharmaceutical, because the methods to manufacture all these have been honed to perfection. Still, while the products of engineers shouldn't clamor for attention by failing, why has the thought of engineers been invisible?

In a search for answers, I absorbed the work of the few writers who were thoughtful on engineers and engineering: Henry Petroski's stunning *The Pencil: A History of Design and Circumstance*, Samuel Florman's revealing *The Existential Pleasures of Engineering*, John Staudenmaier's rich *Technology's Storytellers*, and an early version of Billy Vaughn Koen's audacious *Discussion of the Method*—audacious because he claims the engineering method is *the* universal method, the way to all knowledge. While Koen's book most directly informs this book, it was Florman's *The Introspective Engineer* that firmly laid a foundation of how to think about engineering. In clear, straightforward, and graceful prose deeply rooted in a humanist perspective and the liberal arts, he analyzed why engineers were anonymous, explored the mindset of engineers, and reflected on how best to train engineers.

Buttressed by these foundations, as I neared my tenth year as an engineering professor, I focused my career on explaining engineering to the public, revealing the dazzling creativity of engineers, first on public radio, where I broadcast three- to four-minute pieces about every possible engineering topic. I have never loved a medium as much as I loved radio: it's intellectual (it is, after all, only words!), and it forms a bond with listeners because it typically reaches them when they are alone and listening carefully—in their bathroom in the morning, at breakfast, in their car—and the production is delightfully simple, with no need to shave, shower, or even wear pants. Later, I moved to internet-delivered video—YouTube—where I reached a younger audience.

The impetus to do this was partly that primitive desire to defend my tribe, those engineers who create our world, although that was ultimately more of a justification than a motivation. The appeal to me was that it fused the intellectual inheritance from my parents. My mother, trained as a botanist, delighted in revealing to me and my siblings the natural world. She froze bees so we could study them, drew her own blood for us to examine under a microscope, and, on walks through the nearby woods, identified trees, plants, and animals. On those walks, she transformed what was to me a brown and green patch of land into an evolving, dynamic ecosystem filled with hidden treasures and delights—an analog for revealing the hidden world of engineers. To this day, I enjoy "designing" something so people can understand it; the challenge delights my mind, and what greater challenge than something unknown and unsuspected and hidden like the engineering method.

I combined her powerful example of explanation with my father's passion: theater. As a child, I attended his rehearsals, watched and discussed movies with him, traveled to New York and London to watch plays and musicals with him, and acted in his plays—my brother, sister, and I were the children in the play *JB*, a retelling of the biblical story of Job. Job, of course, loses his family and worldly goods, which relieved my mother: our characters were killed in the early scenes, so we were off to bed on time. My father was no brooding artist; he was a craftsman. His choice of play to direct hewed to the accessible—he liked to entertain. While there was much to learn from him about performance, his greatest impact on me was highlighting the fact that to create something for public consumption, you must hone your craft. A vivid childhood memory is of him preparing for a play. He would lay out a sheet of white paper on which was carefully sketched an overhead view of the set; then, using pennies and dimes to represent the actors, he would work out in detail the blocking, the movements of the actors. I enjoy craft: the shaping of a message, the camera shot that tells, the image that reveals, the words that clarify and enlighten.

Now, in what must be the last phase of my career, I have sifted through fifty years of observing engineering to reveal the deepest roots of how engineers think. To shape that amorphous mass of material, I returned to a childhood influence from another media revolution: the rise of PBS in the late 1960s and early 1970s. The network broadcast Jacob Bronowski's television series *The Ascent of Man*. Just writing the title draws forth a vivid memory of the ominous opening theme as an illustration of a fetus, drawn by Leonardo da Vinci, spun slowly. The show featured Bronowski—a short, wizened man, an unlikely television star!—his gaze fixed firmly on the camera lens, talking directly to the viewer in a conversational tone (there, I now see a seed planted long ago that informs my approach to video!). In broad strokes, ranging over all history and eras, he knit together stories from disparate locales—among them Easter Island, Machu Picchu, Newton's library, Gauss's observatory, and the caves of Altamira—to highlight scientific thought as a pinnacle of human achievement. Bronowski's *The Ascent of Man* still dazzles, but when I returned to it some fifty years later as an inspiration for this book, I looked at his underlying principles in constructing the show. "My ambition," he wrote, was to present "a philosophy rather than a history, and a philosophy of nature rather than of science." At the root was an exploration of humankind: "There cannot be a philosophy, there cannot even be a decent science, without humanity. I hope that sense of affirmation is manifest in this book."[4]

Bronowski's touchstones inform this book. Like Bronowski, I wanted to celebrate these often unheralded, hidden intellectual triumphs that create the world around us. I also wanted not a history but a philosophy in that broad sense of a systematic way to think about the practice of engineering, and I wanted to moor it in what it means to be human. As my career progressed, the human aspects of a technological invocation fascinated me, in contrast to my focus as a student on numbers, on abstraction—an error in our engineering education. For me, technology moved from this implacable force that hems us in and shapes our world to the realization that to succeed, a technology must somehow fit into our world. I thought that only logic dictated solutions, yet I learned that culture influences design. A simple example: the top of a soda can is narrower than the bottom to save aluminum; pure logic suggests an engineer should continue to diminish the size of the top, yet once it becomes too small, consumers reject it because it evokes a container of engine additive.

Yet this just touches the surface. To engineer cuts to the core of being human. Obviously, it fulfills our basic needs of food and shelter, but the impulse to engineer, to create also lies at our core. This impulse is illustrated by one of the earliest civilizations. Over twelve thousand years ago in Turkey, nomadic people built an elaborate temple of limestone, pillars eighteen feet tall, weighing sixteen tons, and with bas-reliefs of snakes, scorpions, boars, and lions, all ready to attack. This structure at Göbekli Tepe (literally "potbelly hill") was dismissed in the 1960s as a mere graveyard but then rediscovered in 1994. What stuns is not merely the construction but the motive; archaeologists discovered no living places nearby, no food debris or shelter, which demonstrates the engineering impulse at its essence: the compulsion to build, to create. As one engineering commentator suggests, "To engineer is human."[5]

So at the most abstract level, I wrote this book to get across one single thought, one takeaway: engineering is a creative profession that surpasses all other human endeavors as a demonstration of the supreme suppleness of the human mind. Yet I also have a personal motive or, better put, personal pleasure: to research and understand these engineering miracles warmly reminds me of my father, a frisson of happy times. Times that I knew were joyful for him because one day, long after our family vacations to factories had ceased, I visited my father, then a widower who lived alone. He looked at me and said, "Would you like to visit a factory?" He had a wistful look, so I said yes. And we trotted off to a nearby chemical factory and sat, father and son, for a quiet hour or two watching plastic wrap being made.

READING GROUP GUIDE

1. *The Things We Make* unpacks the engineering method and the
   ways it has been used throughout human history to build the
   world we know today. After finishing the book, though, what
   are some of the method's implications for the future?
2. Discuss an engineer's definition of "best" and its failings.
   Now, think about an object and identify in what ways it fulfils
   the definition of "best". Conversely, how is it shaped by bias?
   In weighing the positive results of the invention with its flaws,
   do you think that one outweighs the other?
3. The engineering method is, by many counts, an "invisible"
   method, especially when compared with the widely acknowl-
   edged scientific method. Define the differing definitions,
   goals, and outcomes of each.
4. Uncertainty is generally considered a negative trait, but it is
   also a fundamental part of engineering. How do engineers

work around the unknown? In which cases (if any) is uncertainty useful?

5. Describe the three core methods of the engineering mindset and discuss how you have applied them to your own problem solving.

6. There are many ways to define success in engineering. In thinking about the "race to light the world" between Edison, Maxim, and countless other inventors, what factors resulted in Edison being crowned the victor? What factors dictate whether an inventor will be remembered?

7. We are currently facing a number of huge, critical threats. What are they, and how can the engineering method help us face them?

8. Now that you understand the scope of the engineering method, does it change the way you see the things around you? In what way?

APPENDIX

Themes

As this book has demonstrated, the *results* of engineering are not the products of science. The scientific method creates knowledge; the engineering method creates solutions. The products of engineers arise from a method, almost a mindset, that is rarely articulated, almost invisible, yet is universal across cultures and throughout history—no society has survived without some form of the engineering method. The engineering method in action is so novel to most readers, so in this appendix, I recap the main ideas and themes.

At its simplest, the engineering method is *using rules of thumb to solve problems for which there is incomplete information.*

The "problem" of an engineer is their desire to create a device, a process, or a system—a microchip, a chemical plant, or an internet search engine—and the solution, of course, is how to manufacture that microchip, design that plant, or create that algorithm for a search engine. To do this, engineers use rules of thumb, more properly called *heuristics.* These rules of thumb have surprising, even counterintuitive characteristics for a tool that has built civilizations: they don't guarantee a correct answer, and their "correctness" depends on context. An engineer's heuristic method isn't intended to capture any fundamental truths about nature but is a tool that reduces the time for solving a problem.

To create, engineers use three broad classes of rules of thumb:

- *Rules that encapsulate a body of specific knowledge:* This class of rule guides the search for a solution by restricting the search to promising alternatives ("use a factor of safety of 2.0 for bolts in an elevated walkway").
- *Rules that capture unarticulated knowledge:* This class includes knowledge that is passed down through the centuries. These are often "recipes" where the specific reasons for each ingredient or step are unknown.
- *Rules that guide the approach to solving a problem:* Common rules in this class include "at some point in a project, freeze the design," "quantify or express all variables as numbers," and "work at the margins of solvable problems."

With this understanding of the nature of rules of thumb, the core of the engineering method, we can state the method in a more precise way that paves the way for deep understanding of how engineers create: *solving problems using rules of thumb that cause the best change in a poorly understood situation using limited resources.*

Each word in this definition carries a special meaning for an engineer.

An engineer creates the best solution possible, but the word *best* is not used in the commonplace sense of the word, which means the best in an absolute way. This ideal plays *no* part in an engineer's notion of best. The best-engineered design emerges from juggling hundreds of restrictions—cultural forces, societal values, availability of material resources, and even urgency. The best solution is best "all things considered," because an engineer creates within a culture, not in a vacuum.

The engineering meaning of *best* becomes most apparent in a cross-cultural comparison of solutions to the same problem, which reveals that an engineering solution can only be judged based on its response to the constraints unique to a particular society. While there are many reasons for the different solutions in each society, three key aspects stand out:

- Labor supply
- Other technologies available
- Materials readily available

This idea of best being relative to what exists in a society highlights the fact that no technology is neutral: it carries within it the choices and biases of the engineers who created the technology, most often white, middle-class men. Because of this, the products, systems, or algorithms created by engineers often discriminate against women and minorities. This is one of the strongest arguments for a diverse engineering workforce.

The definition of the engineering method includes the phrase *a poorly understood situation*. Although clearly an impediment, this is the central reason that the engineering method exists. To lack information yet design something useful signals that an engineer is at work. In the absence of complete information, engineers for centuries have created buildings, devices, and systems that revolutionized the world. This notion sharply contrasts with a naive view of engineering as "applied science," the idea that first a scientist thoroughly understands a phenomenon, then engineers do some uninteresting development until a product appears. Yet engineers cannot wait for science: bridges, mobile phones, and, more importantly, medicine are needed *today*!

To meet this need, engineers tackle problems where uncertainty pervades; instead of waiting for full understanding, they supplely work their way around uncertainty. They fence it off, then do an end run around it to finalize a design. The key technique is to describe a physical phenomenon of importance to a solution *phenomenologically*, that is, to describe and classify the phenomenon without an explanation or cause—no waiting for a fundamental, perhaps molecular-level, understanding. For some phenomena, engineers use a coarse-grained approach to overcome complexity by "bundling" or "lumping" the phenomenon into a few parameters. This allows an engineer to predict with only a few measurements when a phenomenon will occur. The classic example is the transition from laminar to turbulent flow. Another way to describe something phenomenologically is to use probabilistic methods to predict when something will occur.

These approaches suggest, again, that eventually these phenomena will be fully understood and thus replaced with "science," a loose word for exactitude. It conjures up an image of the tremendous advance and rapid rise of scientific inquiry in the last two hundred years as a rising tide of scientific understanding that will subsume everything. Yet as scientific breakthroughs push out the boundary between the certain and uncertain, engineers move with that boundary, always working slightly beyond the limits of scientific knowledge. This is another reminder that the engineering method is *not* the scientific method.

The definition of the engineering method also contains the phrase *available resources*. As hinted at in looking at the engineering meaning of best and the central role of uncertainty, engineering solutions are bounded by constraints. What an engineer can *use* determines the shape, look, and feel of an engineered product or system. Three key resources are important:

- *Material resources:* At its simplest, this is a truism: if an engineer lived in the Bronze Age, then they used, of course, bronze. This simple but not wrong statement misses the nuance of how engineers weave together available resources in subtle and unexpected ways and how often the *absence* of a resource determines the final look and feel of an object.

- *Energy:* The ingenuity and skill of an engineer in marshaling material resources are readily appreciated, the idea familiar—we all note if a product is plastic or metal. However, the role of energy is often hidden from view while being intimately, intricately, even inseparably intertwined with an engineering design. This constraint is of critical importance when technologies evolve, change, and adapt, and we ignore it to our detriment. For example, by seeing the indivisible link between design and energy, we can nuance and deeply understand the scope and scale of our world's energy challenges. Instead of only thinking of energy in a global, coarse-grained way—those commonplaces of conserving energy, replacing finite energy sources with renewables, ameliorating the effects of climate change—we can realize the profound engineering challenges in finding solutions to our energy woes.

- *Knowledge:* As important as these tangible resources is the tacit engineering knowledge gained over decades of experience. This type of resource highlights the importance of diversity— something mentioned earlier as a way to remove bias—because we need all possible knowledge for engineers to help solve the pressing problems of our time. For many centuries, we have excluded half or more of the population from being engineers; the proper use of this resource supports the idea of diversity in engineering. This is *not* to imply that there is a "female" way of doing engineering but rather to increase the number of *individuals* whose unique knowledge might contribute to a solution.

To apply the engineering method—to create using rules of thumb that cause the best change in a poorly understood situation using available resources—requires the use and embrace of an engineering mindset or attitude reflected in three interrelated, core methods:

- *Trial and error:* At its basis, this is, of course, learning from errors, but to powerfully apply it as part of the engineering method requires deep intuition and meticulous record keeping. This method is not a blind or random search; it is a systematic exploration of a design space, where an engineer varies the value of design parameters within that space. Underlying this are creative and evolving theoretical ideas about what performance characteristics to measure, the possible impact of the parameters, and even which parameters to vary. In these trials, the results are initially unknown, but, guided by institution, the method is not random.
- *Building on past knowledge:* An engineer is essentially a conservative with regard to the use of knowledge—conservative not in the political sense but in the broader sense: "characterized by a tendency to preserve or keep intact or unchanged; preservative," says the *Oxford English Dictionary*.[1] When one is building a bridge, it is better to make small changes in the state of the art; the radical engineer risks building a bridge that fails.
- *Trade-offs:* Any engineered design has limitations, so when designing a product, an engineer must decide how much to

balance the particular characteristics of an object. This item is closely connected with the notion of best. There is never a true, rational answer, nor even an optimal solution, but only a balance among the many constraints.

The engineering method, as has been demonstrated throughout this book, is not the scientific method, yet clearly science and engineering are related. The rise of science over the last 150 years or so has supercharged engineering and created a torrent of new, dazzling technologies that have revolutionized our lives. This leads, as noted earlier, to the naive belief that engineering is applied science—the idea that scientists discover something, and then some (usually perceived as dull) engineering occurs. This pictures science as an organized battlefront that, as it expands, conquers all and delivers technological marvels, yet in reality, engineers fight a guerrilla war. Looking in detail at the development of any piece of technology reveals the following:

- No bit of scientific data or theory explains *how* to design something. That can only be done by imagination, intuition, and superb skills appropriate to the task.
- The role of science is to provide better rules of thumb, specifically scientific data and theories that guide initial choices in a design, and cull out possible solutions from a vast number of possibilities. So science equips engineers with better rules of thumb—rules that eliminate unproductive paths and suggest fruitful ones. Science, then, is the machinery that fuels modern engineering, and to assume engineering is merely applied science is to confuse the machinery of the method with the method itself.
- In a visceral, tangible way, the scientific method creates knowledge, while the engineering method creates solutions.

- Various technological "explosions"—eras with great change—only occurred when science took a step forward and could thus provide better rules of thumb for engineers. Examples are the Bronze Age, the Great Victorian Inheritance (1867–1914), and the quantum physics era (1920s–today).

Mathematics stands in the same relation to engineering as does science. The stereotype of the engineer is that of a geek with a calculator in his pocket protector, which implies that the role of mathematics is to provide the precision needed by engineers, and certainly engineers *do* use mathematics to calculate, size, and specify, but in truth, mathematics plays the same role as science: to provide engineers with powerful rules of thumb. These rules are, of course, not mathematical theorems, and a mathematician would condemn them for their inconsistency and imprecision. The best way to understand the use of mathematics as a powerful rule of thumb is to look at the use of statistics by engineers, which reminds us that engineering is not defined by artifacts (products) but also includes systems and algorithms.

The single most stunning use of mathematics as a rule of thumb is something called extreme value theory—a type of risk assessment used to predict extreme events like winds, floods, and torrential storms so that engineers can design safe buildings. By taking a deep dive into this theory, we can see:

- That rules of thumb used by engineers are not simplistic, mere aphorisms handed down over centuries but can be enormously sophisticated.
- How engineering rules of thumb evolve, often woven out of centuries of thought.
- How engineering rules of thumb are built from the careful but anonymous (at least not famous!) work of ordinary men and

women, which partly decenters the focus on genius as necessary for good engineering.

- How even the most sophisticated method is still context dependent: global climate change might well make rules based on current extreme value theory irrelevant, which highlights again the fact that an engineer's use of mathematics is not, at its deepest, for precision and accuracy.

The engineering method also helps us understand how to think about invention and innovation. More properly, an "invention" is often best regarded as the evolution of design for manufacture. A deep dive into how the manufacturing of a particular product evolves does the following:

- It highlights the myth of the sole inventor, those for whom a spark of inspiration revolutionized the world.
- It reveals how these neat, tidy, digestible sole inventor stories are grossly incomplete and thus distort the true picture, which hides the engineering method; they hide all that is interesting and significant about the engineering method and thus create an unreasonable vision of how innovations occur and unreasonable expectations of engineering.
- It uncenters the movement of invention, reveals the evolution of technological innovation, and lifts from the background previously underappreciated people in that development—often women and people of color—unveiling the previously hidden creativity unleashed by responding to the constraint of designing for mass manufacturing, a constraint beyond that of uncertainty and resources, which often evokes the most ingenious solutions.
- It corrects the distorted view that the root of every new invention is a scientific breakthrough and thus unique; in reality, all

engineering achievements follow the path outlined by the engineering method.

In summary, the engineering method reveals how to *think* about technology and innovation. Understanding the method dispels harmful myths, which arise from four commonplace ways of thinking about technology, the catchall but deficient term used to describe the work of engineers:

- The myth of the lone inventor
- The myth of engineering as "just" applied science
- The myth of a linear trajectory: an engineer seeks a goal and then works toward that goal alone
- The myth of engineering design as objective and separate from the world around it

The impact of this mythical, misbegotten way of seeing engineering results in the following problems:

- We misjudge past engineering achievements and thus discount the power of the engineering method to help us solve today's pressing problems. If we use *today's* state of the art to judge the past, we misread the past, and the result is that we condemn technology in general. This loss of faith in the work of engineers endangers our society at a time when engineering solutions are essential to our planet's survival.
- It dissuades our best and brightest from recognizing engineering as a creative endeavor, when it is this next generation who must solve the dire problems our world faces.
- It causes the public to view technology—a word so all-encompassing that it obscures rather than illuminates—as a great whirlwind that sweeps through our world, creating

blessings and havoc, unrestrained and unable to be controlled, which results in inaction by the public.

- We expect from *scientific* research the groundbreaking, revolutionary inventions of engineers and thus fail to allocate funds for the pure research that generates the powerful rules of thumbs needed by engineers.

Understanding and appreciating the engineering method dispels these myths because one can see the following:

- That we must judge past engineering achievements based on the state of the art—the collection of all heuristics available to an engineer in a particular era, the best engineering practice.
- That to engineer uses the highest degree of humankind's creativity; for millennia now, its results have been the pinnacle of human ingenuity.
- That a successful engineering design reflects all the forces in society, not just technical restrictions but cultural, political, and social as well, and this means engineering designs reflect bias that we must always seek to remove.
- That the word *technology* is a catchall with no distinct meaning—it can mean everything from the practice of engineers to their products to the financial structures that sustain and support it, and by meaning everything, it means nothing and becomes a word that isolates the products of engineering as alien and thus not able to be controlled.
- That technology must be demystified, grounding it in human practice, so that the public can control the technologies that shape our lives so intimately, deeply, and lastingly.

BIBLIOGRAPHY

I have divided my sources into two categories: (a) books, magazines, journals, patents, and interviews, and (b) newspapers. I cite those that I used frequently in shortened form in the notes.

BOOKS, MAGAZINES, JOURNALS, PATENTS, AND INTERVIEWS

al-Hassan, Ahmad Y., and Donald R. Hill. *Islamic Technology: An Illustrated History*. Cambridge, UK: Cambridge University Press, 1986.

al-Jazarī, Ibn al-Razzāz. *The Book of Knowledge of Ingenious Mechanical Devices*. Translated by Donald R. Hill. Dordrecht, Netherlands: D. Reidel, 1974.

Allen, Jack. "The Life and Work of Osborne Reynolds." In *Osborne Reynolds and Engineering Science Today*, edited by D. M.

McDowell and J. D. Jackson, 1–82. Manchester, UK: Manchester University Press, 1970.

Arendt, Hannah. *The Human Condition*. Chicago: University of Chicago Press, 1958.

Arnold, Frances H. "Innovation by Evolution: Bringing New Chemistry to Life (Nobel Lecture)." *Angewandte Chemie International Edition* 58, no. 41 (2019): 14420–26. https://doi.org/10.1002/anie.201907729.

———. "The Library of Maynard-Smith: My Search for Meaning in the Protein Universe." In *Microbes and Evolution: The World That Darwin Never Saw*, edited by R. Kolter and S. Maloy, 203–8. Washington, DC: ASM Press, 2012.

———. "The Nature of Chemical Innovation: New Enzymes by Evolution." *Quarterly Reviews of Biophysics* 48, no. 4 (2015): 404–10. https://doi.org/10.1017/S003358351500013X.

Bahadori, Mehdi N. "Passive Cooling Systems in Iranian Architecture." *Scientific American* 238, no. 2 (February 1978): 144–55. https://www.jstor.org/stable/24955643.

Baker, G. R. M. "Lippmann on Colour Photography." *British Journal of Photography* (April 4, 1896): 265.

Baxter, James Phinney. *Scientists Against Time*. Boston: Little, Brown, 1946.

Bennett, J. H. *Natural Selection, Heredity, and Eugenics, including Selected Correspondence of R. A. Fisher with Leonard Darwin and others*. Oxford, UK: Clarendon Press, 1983.

Berg, Maxine. *Luxury and Pleasure in Eighteenth-Century Britain*. London: Oxford University Press, 2005.

Blank, Paula. *Shakespeare and the Mismeasure of Renaissance Man*. Ithaca, NY: Cornell University Press, 2006.

Bodmer, Walter, R. A. Bailey, Brian Charlesworth, Adam

Eyre-Walker, Vernon Farewell, Andrew Mead, and Stephen Senn. "The Outstanding Scientist, R. A. Fisher: His Views on Eugenics and Race." *Heredity* 126 (2021): 565–76. https://doi.org/10.1038/s41437-020-00394-6.

Boethius. *The Consolation of Philosophy*. Translated by David R. Slavitt. Cambridge, MA: Harvard University Press, 2008.

Borges, Jorge Luis. *The Library of Babel*. Translated by Andrew Hurley. Boston: David R. Godine, 2000.

Borrell, Brendan. "Physics on Two Wheels." *Nature* 535, no. 7612 (July 21, 2016): 338–41. https://www.nature.com/articles/535338a.pdf?origin=ppub.

Bowen, E. G. *Radar Days*. London: Institute of Physics Publishing, 1998.

Bray, Francesca. "How Blind Is Love?: Simon Winchester's 'The Man Who Loved China.'" *Technology and Culture* 51, no. 3 (July 2010): 578–88. https://doi.org/10.1353/tech.2010.0015.

Brayer, Elizabeth. *George Eastman: A Biography*. Rochester, NY: University of Rochester Press, 2006.

Brill, Yvonne Claeys. Dual thrust level monopropellant spacecraft propulsion system. US Patent 3,807,657, filed January 31, 1972, and issued April 30, 1974. https://patents.google.com/patent/US3807657A/en.

———. Interview by Deborah Rice, November 3, 2005. Interview LOH001952.4, Profiles of SWE Pioneers Oral History Project, Walter P. Reuther Library and Archives of Labor and Urban Affairs, Wayne State University, Detroit, MI. https://ethw.org/Oral-History:Yvonne_Brill.

Bronowski, Jacob. *The Ascent of Man*. Boston: Little, Brown, 1973.

Brower, Andrew V. Z. "Fifty Shades of Cladism." *Biology*

and Philosophy 33, no. 8 (2018): 7–11. https://doi. org/10.1007/s10539-018-9622-6.

Bush, Vannevar. *Science—The Endless Frontier: A Report to the President on a Program for Postwar Scientific Research.* Washington, DC: National Science Foundation, 1960. First published 1945.

"Business Notes." *National Engineer* 10, no. 8 (August 1906): 38, 40, 42, 44, 46, 48, 50.

Carneiro, Robert L. "The Evolution of the Tipití: A Study in the Process of Invention." In *Cultural Evolution: Contemporary Viewpoints,* edited by Gary M. Feinman and Linda Manzanilla, 61–93. Dordrecht, Netherlands: Kluwer Academic/Plenum Publishers, 2000.

Chaldecott, John A. "Josiah Wedgwood (1730–95): Scientist." *British Journal for the History of Science* 8, no. 1 (March 1975): 1–16. https://www.jstor.org/stable/4025813.

Chen, K. Q., and F. H. Arnold. "Enzyme Engineering for Nonaqueous Solvents: Random Mutagenesis to Enhance Activity of Subtilisin E in Polar Organic Media." *Biotechnology* 9 (1991): 1073–77. https://doi .org/10.1038/nbt1191-1073.

Clement, Charles R., William M. Denevan, Michael J. Heckenberger, Andre Braga Junqueira, Eduardo G. Neves, Wenceslau G. Teixeira, and William I. Woods. "The Domestication of Amazonia before European Conquest." *Proceedings of the Royal Society B* 282, no. 1812 (2015): 1–9. https://doi.org/10.1098/rspb.2015.0813.

Cohen, Meredith. *The Sainte-Chapelle and the Construction of Sacral Monarchy: Royal Architecture in Thirteenth-Century Paris.* New York: Cambridge University Press, 2015.

Coles, Stuart G., and Elwyn A. Powell. "Bayesian Methods

in Extreme Value Modelling: A Review and New Developments." *International Statistical Review* 64, no. 1 (April 1996): 119–36. https://doi.org/10.2307/1403426.

Collins, Robert M. "History of Agronomy at the Iowa State College." PhD Diss., Iowa State College, 1953. https://dr.lib.iastate.edu/handle/20.500.12876/66686.

"Colour Photography at the Royal Institution." *Photography: The Journal of the Amateur, the Professional, and the Trade* 8, no. 389 (1896): 281.

Coolidge, William D. "Ductile Tungsten." *Transactions of the American Institute of Electrical Engineers* 29, no. 2 (1910): 961–65. https://doi.org/10.1109/T-AIEE.1910.4764659.

———. Tungsten and method of making the same for use as filaments of incandescent electric lamps and for other purposes. US Patent 1,082,933A, filed June 19, 1912, and issued December 30, 1913. https://patents.google.com/patent/US1082933A/en.

Cornish, Paul. *Machine Guns and the Great War*. South Yorkshire, UK: Pen & Sword Military, 2009.

Cowan, Ruth Schwartz. *More Work for Mother: The Ironies of Household Technology from the Open Hearth to the Microwave*. New York: Basic Books, 1983.

Criado Perez, Caroline. *Invisible Women: Data Bias in a World Designed for Men*. New York: Abrams Press, 2019.

Cumming, Robert B. "Is Risk Assessment a Science?," *Risk Analysis* 1, no. 1 (1981): 1–3.

Darrigol, Olivier. *Worlds of Flow: A History of Hydrodynamics from the Bernoullis to Prandtl*. Oxford: Oxford University Press, 2005.

David, F. N. *Games, Gods and Gambling: The Origins and History of Probability and Statistical Ideas from the Earliest Times to the Newtonian Era*. London: C. Griffin, 1962.

Dickinson, H. W. *A Short History of the Steam Engine*. New York: MacMillan, 1958.

Dreyfuss, Henry. *Designing for People*. New York: Allworth Press, 2003.

Eberhart, Jonathan. "The Gentle Rockets." *Science* 91, no. 4 (January 28, 1967): 95, 97. https://doi.org/10.2307/3951521.

Elliott, Gordon. *Aspects of Ceramic History*. 3 vols. Endon, UK: G. W. E. Publications, n.d.

Ewing, J. A. "The Hon. Sir Charles Parsons, O.M., K.C.B. 1854–1931." *Proceedings of the Royal Society of London* Series A 131, no. 818 (June 3, 1931): v–xxv. https://doi.org/10.1098/rspa.1931.0068.

Flint, Charles R. *Memories of an Active Life: Men, and Ships, and Sealing Wax*. New York: G. P. Putnam's Sons, 1923.

Florman, Samuel C. *The Existential Pleasures of Engineering*. New York: St. Martin's Press, 1976.

———. *The Introspective Engineer*. New York: St. Martin's Press, 1996.

"45 Beacon." *Bulletin of the American Meteorological Society* 90, no. 5 (May 2009). https://doi.org/10.1175/1520-0477-90.5.706.

Fouché, Rayvon. *Black Inventors in the Age of Segregation: Granville T. Woods, Lewis H. Latimer, and Shelby J. Davidson*. Baltimore, MD: Johns Hopkins University Press, 2003.

Friedel, Robert, and Paul Israel. *Edison's Electric Light: Biography of an Invention*. Baltimore, MD: Johns Hopkins University Press, 2010.

Gamble, Susan A. "The Hologram and Its Antecedents 1891–1965: The Illusory History of a Three-Dimensional Illusion." PhD Diss., Wolfson College, University of

Cambridge, 2004, https://doi.org/10.17863
/CAM.27481.

Gascoigne, John. "'Getting a Fix': The *Longitude* Phenomenon."
Isis 90, no. 4 (December 2007): 769–78. https://doi
.org/10.1086/529268.

Gladstone, J. H. "Henry Victor Regnault." *Journal of the Chemical
Society* 33 (1878): 235–39, https://doi.org/10.1039
/CT8783300221.

Goldberg, Vicki. "Louise Dahl-Wolfe." *American Photographer* 6,
no. 6 (June 1981): 38–46.

Gorn, Elmer J. "Micro Wave Cooking—The Story of a Man and
His Inventions." Unpublished manuscript, ca. 1970,
Raytheon Archives.

Gould, Stephen Jay. "Ladders and Cones: Constraining Evolution
by Canonical Icons." In *Hidden Histories of Science*, edited
by Robert B. Silvers, 37–67. New York: New York Review
of Books, 1995.

———. "Redrafting the Tree of Life." *Proceedings of the American
Philosophical Society* 141, no. 1 (March 1997): 30–54.
http://www.jstor.org/stable/987248.

———. "The Smoking Gun of Eugenics." *Natural History* 100, no.
12 (December 1991): 8–17.

Gumbel, Emil J. *Statistical Theory of Extreme Values and Some
Practical Applications: A Series of Lectures*. Applied
Mathematics Series 33. Washington, DC: National Bureau
of Standards, U.S. Department of Commerce, 1954.

Heimpel, Wolfgang. *Letters to the King of Mari: A New Translation,
with Historical Introduction, Notes, and Commentary*.
Winona Lake, IN: Eisenbrauns, 2003.

Herlihy, David V. *Bicycle: The History*. New Haven, CT: Yale
University Press, 2004.

Heyman, J. "On the Rubber Vaults of the Middle Ages and Other Matters." In *The Engineering of Medieval Cathedrals*, edited by Lynn T. Courtenay, 15–26. Brookfield, VT: Ashgate, 1997.

Hill, Donald R. *Islamic Science and Engineering*. Edinburgh, UK: Edinburgh University Press, 1993.

Huerta, Santiago. "Galileo Was Wrong: The Geometrical Design of Masonry Arches." *Nexus Network Journal* 8, no. 2 (2006): 25–52. https://doi.org/10.1007/s00004-006-0016-8.

———. "Geometry and Equilibrium: The Gothic Theory of Structural Design." *Structural Engineer* 84, no. 2 (January 17, 2006): 23–28.

Hughes, Thomas P. *American Genesis: A Century of Invention and Technological Enthusiasm, 1870–1970*. Chicago: University of Chicago Press, 2004.

Hutchinson, Alex. *Big Ideas: 100 Modern Inventions That Have Transformed Our World*. New York: Hearst Books, 2009.

Hyde, Ralph. "Prints and Wedgwood." *Print Quarterly* 13, no. 2 (1996): 208–11. https://www.jstor.org/stable/41825291.

Iowa State University Faculty Senate. "Memorial Resolutions," May 5, 2009. https://www.facsen.iastate.edu/sites/default/files/uploads/Memorial%20Resolutions/Memorial%20Resolutions20090505.pdf.

Ives, Herbert E. "An Experimental Study of the Lippmann Color Photograph." *Astrophysical Journal* 27 (1908): 325–52.

Jehl, Francis. *Menlo Park Reminiscences*. 3 vols. Dearborn, MI: Edison's Institute, 1936–1941.

Johnson, Bob. *Shuji Nakamura and the Revolution in Lightning Technology*. Amherst, NY: Prometheus Books, 2015.

Kevles, Daniel J. *In the Name of Eugenics: Genetics and the Uses of Human Heredity*. New York: Knopf, 1985.

Knell, Simon. "A Palatable Myth of William Smith." *Metascience* 11, no. 2 (June 2002): 261–65.

Koen, Billy Vaughn. *Discussion of the Method: Conducting the Engineer's Approach to Problem Solving*. Oxford: Oxford University Press, 2003.

Kooijman, J. D. G., J. P. Meijaard, Jim M. Papadopoulos, Andy Ruina, and A. L. Schwab. "A Bicycle Can Be Self-Stable Without Gyroscopic or Caster Effects." *Science* 332, no. 6027 (April 15, 2011): 339–42. https://doi.org/10.1126/science.1201959.

Kramer, Edna E. *The Nature and Growth of Modern Mathematics*. Princeton, NJ: Princeton University Press, 1970.

Krause, Peter. "50 Years of Kodachrome." *Modern Photography* 49, no. 11 (October 1985): 47–63, 83, 94, 96, 98, 104, 106, 112, 114.

Kruft, Hanno-Walter. *History of Architectural Theory*. Princeton, NJ: Princeton Architectural Press, 1994.

Kurzweil, Ray. *The Singularity Is Near: When Humans Transcend Biology*. New York: Viking, 2005.

Laskow, Sarah. "Inventing the LED Lightbulb." *The Atlantic*, September 10, 2014. https://www.theatlantic.com/technology/archive/2014/09/who-invented-the-new-lightbulb/379905/.

Latimer, Lewis H. Process of manufacturing carbons. US Patent 252,386, filed February 19, 1881, issued January 17, 1882. https://patents.google.com/patent/US252386A/en.

Leggett, Don. "Spectacle and Witnessing: Constructing Readings of Charles Parsons's Marine Turbine." *Technology and Culture* 52, no. 2 (2011): 287–309. https://doi.org/10.1353/tech.2011.0043.

Lemley, Mark A. "The Myth of the Sole Inventor." *Michigan Law*

Review 110, no. 5 (2012): 709–60. https://doi
.org/10.2139/ssrn.1856610.

Lévi-Strauss, Claude. *Tristes Tropiques*. Translated by John and
Doreen Weightman. New York: Penguin, 1992.

Leyden, Dennis Patrick, and Matthias Menter. "The Legacy and
Promise of Vannevar Bush: Rethinking the Model of
Innovation and the Role of Public Policy." *Economics of
Innovation and New Technology* 27, no. 3 (2018): 225–42.
https://doi.org/10.1080/10438599.2017.1329189.

Light, Jennifer S. "When Computers Were Women." *Technology
and Culture* 40, no. 3 (1999): 455–83. http://www.jstor
.org/stable/25147356.

Long, Pamela O. *Technology and Society in the Medieval Centuries:
Byzantium, Islam, and the West, 500–1300.* Washington,
DC: American Historical Association, 2003.

Louçã, Francisco. "Emancipation Through Interaction—How
Eugenics and Statistics Converged and Diverged." *Journal
of the History of Biology* 42, no. 4 (Winter 2009): 649–84.
https://doi.org/10.1007/s10739-008-9167-7.

Lyman, Frederic A. "A Practical Hero: Or, How an Obscure New
York Mechanic Got a Steam-Powered Toy to Drive
Sawmills." *Mechanical Engineering* 126, no. 2 (February
2004): 36–38.

MacAdam, David L. "Quality of Color Reproduction." *Journal of
the Society of Motion Picture and Television Engineers* 56
(May 1951): 487–512. https://doi.org/10.5594/J06314.

MacKenzie, Donald A. *Statistics in Britain 1865–1930.* Edinburgh,
UK: Edinburgh University Press, 1981.

Marcum, J., and T. P. Kinn. "Heating with Microwaves." *Electronics*
20 (March 1947): 82–85.

Marrison, Warren A. "The Evolution of the Quartz Crystal Clock."

Bell System Technical Journal 27, no. 3 (July 1948): 510–88. https://doi.org/10.1002/j.1538-7305.1948.tb01343.x.

Marsden, Ben. "Ranking Rankine: W. J. M. Rankine (1820–72) and the Making of 'Engineering Science' Revisited." *History of Science* 51, no. 4 (2013): 434–56. https://doi.org/10.1177/007327531305100403.

Marshall, R. D., and H. C. S. Thom, eds. *Proceedings of Technical Meeting Concerning Wind Loads on Buildings and Structures.* Building Science Series 30. Washington, DC: National Bureau of Standards, U.S. Department of Commerce, 1970.

Maxim, Hiram S. Improvements in devices for advertising purposes. GB Patent 190702482A, filed January 31, 1907, and issued April 30, 1908. https://patents.google.com/patent/GB190702482A/.

———. *My Life.* London: Methuen, 1915. "Maxim's Electric Light." *Engineering* 31 (June 3, 1881): 569–70. "The Maxim System of Electric Illumination by Incandescence." *Engineering* 31 (June 17, 1881): 618–20.

McKendrick, Neil. "Josiah Wedgwood and Factory Discipline." *Historical Journal* 4, no. 1 (1961): 30–55. https://www.jstor.org/stable/3020380.

———. "Josiah Wedgwood and Thomas Bentley: An Inventor-Entrepreneur Partnership in the Industrial Revolution." *Transactions of the Royal Historical Society* 14 (1964): 1–33. https://doi.org/10.2307/3678942.

———. "The Role of Science in the Industrial Revolution: A Study of Josiah Wedgwood as a Scientist and Industrial Chemist." In *Changing Perspectives in the History of Science: Essays in Honor of Joseph Needham*, edited by Mikuláš Teich and Robert Young, 274–319. London: Heinemann, 1973.

Metcalfe, Robert M. Interviews by Len Shustek, November 29, 2006, January 31, 2007. Interview X3819.2007, Computer History Museum, Mountain View, California. https://www.computerhistory.org/collections/catalog/102657995.

Metcalfe, Robert M., and David R. Boggs. "Ethernet: Distributed Packet Switching for Local Computer Networks." *Communications of the ACM* 19, no. 7 (July 1976): 395–404. https://doi.org/10.1145/360248.360253.

Meteyard, Eliza. *The Life of Josiah Wedgwood.* 2 vols. London: Hurst and Blackett, 1866.

Miller, David Philip. "The 'Sobel Effect.'" *Metascience* 11, no. 2 (June 2002): 185–200. https://doi.org/10.1007/BF02914819.

Miller, John Anderson. *Yankee Scientist: William David Coolidge.* Schenectady, NY: Mohawk Development Service, 1963.

Moffett, Cleveland. "The Fastest Vessel Afloat." *Pearson's Magazine* 6 (July–December 1898): 224–30.

Mühlethaler, Bruno, and Jean Thissen. "Smalt." *Studies in Conservation*, no. 2 (1969): 47–61. https://doi.org/10.1179/sic.1969.005.

Mullin, T. "Experimental Studies of Transition to Turbulence in a Pipe." *Annual Review of Fluid Mechanics* 43 (2011): 1–24. https://doi.org/10.1146/annurev-fluid-122109-160652.

Mumford, Lewis. *Technics and Civilization.* Chicago: University of Chicago Press, 1934.

Murray, Don. "Percy Spencer and His Itch to Know." *Reader's Digest* (August 1958): 114–18.

Nakamura, Shuji, Stephen Pearton, and Gerhard Fasol. *The Blue Laser Diode: The Complete Story.* Berlin: Springer-Verlag, 2000.

Neale, B. T. "CH—The First Operational Radar." *GEC Journal of Research* 3, no. 2 (1985): 73–83. https://marconiradarhistory.pbworks.com/f/CH-The%20First%20Operational%20Radar.pdf.

Nichols, Joseph V., and Lewis H. Latimer. Electric lamp. US Patent 247,097, filed April 18, 1881, and issued September 13, 1881. https://patents.google.com/patent/US247097A/en.

Nickerson, Angela K. *A Journey into Michelangelo's Rome.* Berkeley, CA: Roaring Forties Press, 2008.

"Niels Anton Christensen." *Successful American* 8, no. 1 (February 1903): 412–16.

Nobel Lectures Including Presentation Speeches and Laureates' Biographies: Physics 1901–1921. Amsterdam: Elsevier, 1967.

Noble, Safiya Umoja. *Algorithms of Oppression: How Search Engines Reinforce Racism.* New York: New York University Press, 2018.

Null, Roberta. *Universal Design: Principles and Models.* Boca Raton, FL: CRC Press, 2014.

Ore, Oystein. "Pascal and the Invention of Probability Theory." *American Mathematical Monthly* 67, no. 5 (May 1960): 409–19. https://doi.org/10.2307/2309286.

Osepchuk, John M. "A History of Microwave Heating Applications." *IEEE Transactions on Microwave Theory and Techniques* MTT-32, no. 9 (September 1984): 1200–24. https://doi.org/10.1109/TMTT.1984.1132831.

Parsons, Charles A. "The Application of the Compound Steam Turbine to the Purpose of Marine Propulsion." *Transactions of the Institution of Naval Architects* 38 (1897): 232–42.

———. *The Steam Turbine: The Rede Lecture 1911*. Cambridge, UK: Cambridge University Press, 1911.

Pascal, Blaise. *Pensées*. Translated by A. J. Krailsheimer. London: Penguin Books, 1995.

Petroski, Henry. *The Pencil: A History of Design and Circumstance*. New York: Knopf, 1990.

———. *To Engineer Is Human: The Role of Failure in Successful Design*. New York: St. Martin's Press, 1985.

Piperno, Dolores R., Crystal McMichael, and Mark B. Bush. "Amazonia and the Anthropocene: What Was the Spatial Extent and Intensity of Human Landscape Modification in the Amazon Basin at the End of Prehistory?" *Holocene* 25, no. 10 (2015): 1588–97. https://doi.org/10.1177/0959683615588374.

Pope, Franklin Leonard. *Evolution of the Electric Incandescent Lamp*. Elizabeth, NJ: Henry Cook, 1889.

Prak, Maarten. "Mega-Structures of the Middle Ages: The Construction of Religious Buildings in Europe and Asia, c.1000–1500." *Journal of Global History* 6 (2011): 381–406. https://doi.org/10.1017/S1740022811000386.

Quinn, Roswell. "Rethinking Antibiotic Research and Development: World War II and the Penicillin Collaborative." *American Journal of Public Health* 103, no. 3 (March 2013): 426–34. https://doi.org/10.2105/AJPH.2012.300693.

Reilly, Robin. "Josiah Wedgwood, A Lifetime of Achievement." In *Genius of Wedgwood*, edited by Hilary Young, 44–57. London: Victoria and Albert Museum, 1995.

———. *Wedgwood*. 2 vols. New York: Stockton Press, 1989.

———. *Wedgwood Jasper*. London: Thames and Hudson, 1994.

Reynolds, Osborne. *Papers on Mechanical and Physical Subjects*.

3 vols. Cambridge, UK: Cambridge University Press, 1900–3.

"The Rise of American Pay-TV." *The Economist* 268, no. 7039 (July 29, 1978): 62.

"The Rise of Raytheon." *Fortune* 34, no. 6 (October 1946): 136–88.

Rolston, Nicholas, William J. Scheideler, Austin C. Flick, Justin P. Chen, Hannah Elmaraghi, Andrew Sleugh, Oliver Zhao, Michael Woodhouse, and Reinhold H. Dauskardt. "Rapid Open-Air Fabrication of Perovskite Solar Modules." *Joule* 4 (December 16, 2020): 2675–92. https://doi.org/10.1016/j.joule.2020.11.001.

Rose, Hilary. *Love, Power and Knowledge: Towards a Feminist Transformation of the Sciences.* Cambridge, UK: Polity Press, 1994.

Roth, Lorna. "Looking at Shirley, the Ultimate Norm: Colour Balance, Image Technologies, and Cognitive Equity." *Canadian Journal of Communications* 34, no. 1 (2009): 111–36. https://doi.org/10.22230/cjc.2009v34n1a2196.

Rushkoff, Douglas. *Program or Be Programmed: Ten Commands for a Digital Age.* Berkeley, CA: Soft Skull Press, 2011.

Russell, Bertrand. *Our Knowledge of the External World as a Field for Scientific Method in Philosophy.* Chicago: Open Court, 1914.

Salvadori, Mario. *Why Buildings Stand Up: The Strength of Architecture.* New York: W. W. Norton, 1980.

Scaife, W. Garrett. *From Galaxies to Turbines: Science, Technology and the Parsons Family.* Bristol, UK: Institute of Physics, 2000.

———. "The Parsons Steam Turbine." *Scientific American* 252, no. 4 (April 1985): 132–39. http://www.jstor.org/stable/24967620.

Schuster, Arthur. *Biographical Fragments*. London: MacMillan, 1932.

Segal, Howard. *Technological Utopianism in American Culture*. Chicago: University of Chicago Press, 1985.

Sennett, Richard. *The Craftsman*. New Haven, CT: Yale University Press, 2008.

Service, Robert F. "Protein Evolution Earns Chemistry Nobel." *Science* 362, no. 641 (October 12, 2018): 142. https://doi.org/10.1126/science.362.6411.142.

Shelby, Lon R. "The Education of Medieval English Master Masons." *Medieval Studies* 32 (1970): 1–26.

———. "The Geometrical Knowledge of Mediaeval Master Masons." *Speculum* 47, no. 3 (1972): 395–421. https://doi.org/10.2307/2856152.

———. "Mediaeval Masons' Templates." *Journal of the Society of Architectural Historians* 30, no. 2 (1971): 140–54. https://doi.org/10.2307/988630.

———. "Medieval Masons' Tools. II. Compass and Square." *Technology and Culture* 6, no. 2 (1965): 236–48. https://doi.org/10.2307/3101076.

———. "The Role of the Master Mason in Mediaeval English Building." *Speculum* 39, no. 3 (1964): 387–403. https://doi.org/10.2307/2852495.

Shelby, Lon R., and Robert Mark. "Late Gothic Structural Design in the 'Instructions' of Lorenz Lechler." *Architectura* 9, no. 2 (1979): 113–31.

Silver, Daniel S. "Knot Theory's Odd Origins." *American Scientist* 94, no. 2 (March–April 2006): 158–65. https://doi.org/10.1511/2006.58.158.

Simon, Julian L. *The Ultimate Resource 2*. Princeton, NJ: Princeton University Press, 1996.

Singh, Simon. *Fermat's Last Theorem*. New York: HarperCollins, 2012.

Smiles, Samuel. *Josiah Wedgwood, F.R.S., His Personal History*. New York: Harper & Brothers, 1895.

Smil, Vaclav. *Creating the Twentieth Century: Technical Innovations of 1867–1914 and Their Lasting Impact*. Oxford: Oxford University Press, 2005.

Smith, Cyril Stanley. *A Search for Structure: Selected Essays on Science, Art, and History*. Cambridge, MA: MIT Press, 1981.

Smith, Frank E. "Sir Charles Parsons and Steam." *Transactions of the North-East Coast Institution of Engineers and Shipbuilders* 53 (1936–1937): 31–52.

Smoryński, Craig. *Mathematical Problems: An Essay on Their Nature and Importance*. Cham, Switzerland: Spring Nature, 2020.

"The Society's Awards." *Bulletin of the American Meteorological Society* 47, no. 8 (1966): 624–33. https://doi.org/10.1175/1520-0477-47.8.624.

Spencer, Percy L. Method of treating foodstuffs. US Patent 2,495,429, filed October 8, 1945, issued January 24, 1950. https://patents.google.com/patent/US2495429A/en.

———. "P. L. Spencer: Raytheon November 15, 1925–June 1, 1959." Unpublished manuscript, January 25, 1960, Raytheon Archives.

"Spirit of the Times." *Photography: The Journal of the Amateur, the Professional, and the Trade* 8, no. 389 (1896): 272.

Staudenmaier, John M. *Technology's Storytellers: Reweaving the Human Fabric*. Cambridge: MIT Press, 1985.

Swinburne, James. "Incandescent Electric Lamps: Part I." *Technics* 1, no. 2 (February 1904): 160–64.

Tennesen, Michael. "Uncovering the Arawaks." *Archaeology* 63, no. 5 (September/October 2010): 51–52, 54, 56. https://www.jstor.org/stable/41780608.

Thom, H. C. S. "Frequency of Maximum Winds." *Proceedings of the American Society of Civil Engineers* 80, no. 539 (November 1954): 1–11.

Tippett, L. H. C. "Some Applications of Statistical Methods to the Study of Variation of Quality in the Production of Cotton Yarn." *Supplement to the Journal of the Royal Statistical Society* 2, no. 1 (1935): 27–62. https://doi.org/10.2307/2983586.

Turnbull, David. "The Ad Hoc Collective Work of Building Gothic Cathedrals with Templates, String, and Geometry." *Science, Technology, & Human Values* 18, no. 3 (1993): 315–40. https://doi.org/10.1177/016224399301800304.

van Loon, A., P. Noble, D. de Man, M. Alfeld, T. Callewaert, G. Van der Snickt, K. Jansens, and J. Dik. "The role of smalt in complex pigment mixtures in Rembrandt's Homer 1663: combining MA-XRF imaging, microanalysis, paint reconstructions and OCT." *Heritage Science* 8, no. 90 (2020): Herit Sci 8, 90 (2020). https://doi.org/10.1186/s40494-020-00429-5.

Warrillow, E. J. D. *History of Etruria: Staffordshire, England 1760–1951.* 3rd ed. Stoke-on-Trent, UK: Etruscan Publications, 1953.

Wedgwood, Josiah. *Letters of Josiah Wedgwood.* Edited by Katherine Eufemia Farrer. 3 vols. London: Women's Printing Society, 1903–1906. Reprinted Manchester, UK: E. J. Morten for the Trustees of the Wedgwood Museum, 1973.

Wilson, H. W. "The Global Review at Spithead: A Superb Display." *The Graphic*, no. 1440 (July 3, 1897).

Wilson, S. S. "Bicycle Technology." *Scientific American* 228, no. 3

(March 1973): 81–91. https://www.jstor.org
/stable/24923004.

Winston, Brian. "A Whole Technology of Dyeing: A Note on
Ideology and the Apparatus of the Chromatic Moving
Image." *Daedalus* 114, no. 4 (1985): 105–23. https://
www.jstor.org/stable/20025012.

Wise, George. "Ring Master." *American Heritage of Invention
and Technology* 7, no.1 (Spring/Summer 1991): 58–63.
https://www.inventionandtech.com/content
/ring-master-1.

Wiseman, D. J. "Mesopotamian Gardens." *Anatolian Studies* 33
(1983): 137–44. https://doi.org/10.2307/3642702.

NEWSPAPERS

These articles are organized alphabetically by newspaper and then
chronologically for each title.

Liverpool Mercury
"After the Spithead Review: The Cruise of the Teutonic." June
29, 1897.

New York Times
Martin, Douglas. "Yvonne Brill, a Pioneering Rocket Scientist,
Dies at 88." March 30, 2013. https://www.nytimes
.com/2013/03/31/science/space/yvonne-brill-rocket-
scientist-dies-at-88.html.

Times (London)
Advertisement. April 17, 1896, 12.
"The Turbinia." June 28, 1897.

IMAGE CREDITS

Section of a hieroglyph: from Lepsius, Carl Richard. *Denkmäler aus Aegypten und Aethiopien*. Plates 2, Band 3 [Giza plates only]. Berlin: Nicolaische Buchhandlung, 1849–1859.

A kelek: from https://digitalcollections.nypl.org/ RLIN /OCLC: 6801167; NYPL catalog ID (B-number): b14308515; Universal Unique Identifier (UUID): f1a4e960-c6d3–012f-b5b0–58d385a7bc34.

Avery's steam-powered buzz saw: adapted from US Patent 6766X A. Foster and W. Avery "Steam Engine" September 28, 1831.

Wind map: from *Minimum Design Loads and Associated Criteria for Buildings and Other Structures* published by the American Society of Civil Engineers. Used with permission from ASCE.

Maxim's light bulb: from Alglave, Em, and J. Boulard. *The Electric Light: Its History, Production, and Applications*. New York: D. Appleton, 1884, 180.

NOTES

INTRODUCTION

[1] Cohen, *Sainte-Chapelle*, 231.

[2] Turnbull, "Ad Hoc Collective Work," 316.

CHAPTER 1

[1] The most detailed study of what masons knew and how they worked is the lifetime of work by Lon Shelby. See the bibliography for a list of works.

[2] Prak, "Mega-Structures of the Middle Ages," 388.

[3] Turnbull, "Ad Hoc Collective Work," 331.

[4] Shelby and Mark, "Late Gothic Structural Design," 115.

[5] Nickerson, *Journey into Michelangelo's Rome*, 118.

[6] The most detailed yet accessible study of the rules of thumb used by medieval masons is the work of Santiago Huerta. See the bibliography for a list of works.

[7] Shelby and Mark, "Late Gothic Structural Design," 115.

[8] Koen, *Discussion of the Method*, 34.

[9] *Oxford English Dictionary*, s.v. "heuristic," accessed February 04, 2021, https://www.oed.com/viewdictionaryentry/Entry/86554.

10   Heyman, "On the Rubber Vaults," 6.

11   Details of Christensen's life and work can be found in "Niels Anton
      Christensen"; "Business Notes," 44; and Wise, "Ring Master."

12   "Business Notes," 44.

13   Wise, "Ring Master," 61.

14   Wise, "Ring Master," 61.

15   Russell, *Our Knowledge*, 236.

CHAPTER 2

1   Blank, *Shakespeare and the Mismeasure*, 15.

2   Dreyfuss, *Designing for People*, 23–24.

3   Dreyfuss, *Designing for People*, 105.

4   Tennesen, "Uncovering the Arawaks," 51.

5   The extent and complexity of Amazonia are the subject of much debate,
      good entry points to which are Piperno, McMichael, and Bush, "Amazonia
      and the Anthropocene"; and Clement et al., "Domestication of Amazonia."

6   Carneiro, "Evolution of the Tipití," 67.

7   The most detailed study of the tipití is Carneiro, "Evolution of the Tipití."

8   Lévi-Strauss, *Tristes Tropiques*, 215.

9   Crash test dummies and other examples from Criado Perez, *Invisible Women*.

10   Noble, *Algorithms of Oppression*, 64.

11   Advertisement in *Times* (London), April 17, 1896.

12   Gamble, "Hologram and Its Antecedents," 32.

13   Gamble, "Hologram and Its Antecedents," 40.

14   Quotes from Lippman and description of his talk are from "Colour
      Photography"; Baker, "Lippmann on Colour Photography"; and Lippmann's
      description of his work in his Nobel Lecture of December 14, 1908,
      reprinted in *Nobel Lectures*.

15   Ives, "Experimental Study," 326–27.

16   "Spirit of the Times."

17   Brayer, *George Eastman: A Biography*, 220.

18   Brayer, *George Eastman: A Biography*, 205.

19   The story of Godowsky, Mannes, and Kodachrome is well told in Krause, "50
      Years of Kodachrome."

20   Brayer, *George Eastman: A Biography*, 226.

21   *National Geographic*, June 1985.

22   "Final Kodachrome Images Donated to George Eastman House International
      Museum of Photography and Film," Museum Publicity, June 17, 2011,

https://museumpublicity.com/2011/06/17/final-kodachrome-images-donated-to-george-eastman-house-international-museum-of-photography-and-film/.

23   Krause, "50 Years of Kodachrome," 63.

24   A fine history of color reproduction and its inability to reproduce all skin tones is Winston, "Whole Technology of Dyeing."

25   The quotes in this paragraph are from *Kodak Color Dataguide*, 3rd ed. (Rochester, NY: Eastman Kodak, 1968). These instructions are on the envelope containing the Shirley.

26   For a fascinating history of the Shirleys, see Roth, "Looking at Shirley."

27   MacAdam, "Quality of Color Reproduction," 502.

28   Goldberg, "Louise Dahl-Wolfe," 46.

29   All quotes from Georgena Terry are from a discussion with the author, April 24, 2002.

30   For an overview of this subject, see Null, *Universal Design*.

CHAPTER 3

1   Herlihy, *Bicycle: The History*, 18.

2   Wilson, "Bicycle Technology."

3   Borrell, "Physics on Two Wheels," 339.

4   Kooijman et al., "Bicycle Can Be Self-Stable."

5   Kooijman et al., "Bicycle Can Be Self-Stable," 342.

6   Mullin, "Experimental Studies," 1.

7   Sir Graham Sutton as quoted in Allen, "Life and Work," 65.

8   Schuster, *Biographical Fragments,* 228.

9   Reynolds, *Papers*, 2:52.

10   Reynolds, *Papers*, 2:159.

11   Reynolds as quoted in Darrigol, *Worlds of Flow*, 258.

12   Reynolds, *Papers*, 2:58.

13   Reynolds, *Papers*, 1:198.

14   Reynolds, *Papers*, 1:199.

15   Reynolds, *Papers*, 1:185.

16   Reynolds, *Papers*, 2:52.

17   Reynolds, *Papers*, 1:186.

18   For a fascinating account of the smoke ring box and its use by nineteenth-century physicists to answer fundamental questions about the universe, see Silver, "Knot Theory's Odd Origins."

19   Reynolds, *Papers*, 1:186.

[20] Reynolds, *Papers*, 1:187.

[21] Reynolds, *Papers*, 1:188.

[22] Reynolds, *Papers*, 2:70.

[23] Reynolds, *Papers*, 2:72.

[24] Reynolds, *Papers*, 2:158.

[25] Quotes in this paragraph from are Reynolds, *Papers*, 2:155.

[26] Metcalfe and Boggs, "Ethernet," 431.

[27] Dan Pitt (IBM engineer), "IBM's LAN Architecture & Philosophy," History of Wired LAN Competition, December 22, 2014, YouTube video, 5:01, https://www.youtube.com/watch?v=qvCLvqXSmd0.

[28] Pitt, "IBM's LAN Architecture."

[29] Robert Metcalfe, interview by Len Shustek, November 29, 2006, January 31, 2007.

[30] Geoff Thompson (Xerox engineer), "Aloha and then Ethernet," History of Wired LAN Competition, December 22, 2014, YouTube video, 2:35, https://www.youtube.com/watch?v=V5GzgeQCzrk.

[31] A marvelous book on the Great Victorian Inheritance is Smil, *Creating the Twentieth Century*.

[32] All quotes in this paragraph and the next from Arnold, "Library of Maynard-Smith."

[33] Borges, *Library of Babel*, 22, 26.

[34] Arnold, "Innovation by Evolution," 14421.

[35] Arnold, "Innovation by Evolution," 14420.

[36] Service, "Protein Evolution," 142.

[37] Arnold, "Innovation by Evolution," 14422.

[38] Arnold, "Library of Maynard Smith," 207.

[39] Arnold reported her creation of the first enzyme engineered by directed evolution in Chen and Arnold, "Enzyme Engineering."

[40] Arnold, "Nature of Chemical Innovation," 404.

[41] Arnold, "Innovation by Evolution," 14425.

[42] Arnold, "Innovation by Evolution," 14423.

CHAPTER 4

[1] Wiseman, "Mesopotamian Gardens," 142.

[2] For the fascinating details of this cooling system, see Bahadori, "Passive Cooling."

[3] The details and quotations from Sidqum-Lanasi's interactions with Zimri-Lim are from Heimpel, *Letters to the King*.

4   For detailed studies of Islamic engineering, see al-Hassan and Hill, *Islamic Technology*; Hill, *Islamic Science and Engineering*; and Long, *Technology and Society*.

5   For details of early Islamic timekeeping, see Hill, *Islamic Science and Engineering*, 38–39.

6   al-Jazarī, *Book of Knowledge*, 15.

7   For a superb discussion of timekeeping prior to atomic clocks, see Marrison, "Evolution."

8   al-Jazarī, *Book of Knowledge*, 83.

9   Eberhart, "Gentle Rockets."

10   All quotes from Brill in this chapter are from Yvonne Brill, interview by Deborah Rice, November 3, 2005.

11   Rose, *Love, Power and Knowledge*, 132.

12   Light, "When Computers Were Women," 459.

13   "Rise of American Pay-TV."

14   The key patent is Brill, dual thrust level monopropellant spacecraft propulsion system, US 3,807,657.

15   Eberhart, "Gentle Rockets," 95.

16   National Sicence & Technology Medals Foundation, "Yvonne Brill - 2010 National Medal of Technology & Innovation." YouTube video, 2:04. Posted November 3, 2011, https://www.youtube.com/watch?v=6LP2Ni0c1Bg.

17   Martin, "Yvonne Brill."

18   Brill's obituary fails the "Finkbeiner Test," a list of statements developed by journalist Christie Aschwanden to root out gender bias in stories about a woman who is a scientist, mathematician, or engineer. The seven statements that should not appear are as follows: (1) the fact that she's a woman, (2) her husband's job, (3) her child-care arrangements, (4) how she nurtures her underlings, (5) how she was taken aback by the competitiveness in her field, (6) how she's such a role model for other women, and (7) how she's the "first woman to..." The reader might ask whether my profile of Brill passes this test.

CHAPTER 5

1   Reilly, *Wedgwood Jasper*, 80.

2   Chaldecott, "Josiah Wedgwood," 16.

3   McKendrick, "Role of Science," 285, 292.

4   Elliott, *Aspects of Ceramic History*, 3:24.

5   Reilly, "Josiah Wedgwood," 44.

6   Berg, *Luxury and Pleasure*, 132–33.

7   Reilly, "Josiah Wedgwood," 45.

8   Reilly, "Josiah Wedgwood," 45.

9   Reilly, *Wedgwood Jasper*, 74.

10   Reilly, *Wedgwood*, 1:183.

11   Reilly, *Wedgwood Jasper*, 73.

12   Reilly, *Wedgwood Jasper*, 80.

13   McKendrick, "Josiah Wedgwood and Thomas Bentley," 11.

14   Smiles, *Josiah Wedgwood*, 92–93.

15   Meteyard, *Life of Josiah Wedgwood*, 1:486.

16   McKendrick, "Josiah Wedgwood and Thomas Bentley," 13.

17   Typical events of interest to a Londoner as reported in *Bingley's London Journal*, December 12, 1772.

18   For a discussion of Hamilton's books and their impact on British culture in the eighteenth century, see Kruft, *History of Architectural Theory*, 215.

19   Reilly, *Wedgwood Jasper*, 69–70.

20   Reilly, *Wedgwood Jasper*, 70.

21   Wedgwood, *Letters of Josiah Wedgwood*, 2:126. "Double hazardous" was a term of art in eighteenth-century insurance: timber building were "hazardous," while thatched buildings were "double hazardous."

22   Wedgwood, *Letters of Josiah Wedgwood*, 2:189.

23   Wedgwood, *Letters of Josiah Wedgwood*, 2:105.

24   See Warrillow, *History of Etruria* for details of this building.

25   Wedgwood, *Letters of Josiah Wedgwood*, 2:123.

26   Wedgwood, *Letters of Josiah Wedgwood*, 2:106.

27   Reilly, *Wedgwood Jasper*, 74.

28   Reilly, *Wedgwood*, 1:271.

29   Hyde, "Prints and Wedgwood," 208.

30   McKendrick, "Josiah Wedgwood and Factory Discipline," 44.

31   Reilly, *Wedgwood Jasper*, 74.

32   Reilly, *Wedgwood Jasper*, 79.

33   McKendrick, "Josiah Wedgwood and Factory Discipline," 47.

34   Reilly, *Wedgwood*, 1:185.

35   Reilly, *Wedgwood Jasper*, 74.

36   Reilly, *Wedgwood Jasper*, 74.

37   Reilly, *Wedgwood Jasper*, 75.

38   Reilly, *Wedgwood Jasper*, 75.

39   Wedgwood, *Letters of Josiah Wedgwood*, 2:154.
40   Wedgwood, *Letters of Josiah Wedgwood*, 2:155.
41   Reilly, *Wedgwood Jasper*, 74.
42   Reilly, *Wedgwood Jasper*, 75.
43   van Loon, "Role of Smalt."
44   Reilly, *Wedgwood Jasper*, 85.
45   Reilly, *Wedgwood Jasper*, 76, 79.
46   Reilly, *Wedgwood Jasper*, 79.
47   Reilly, *Wedgwood Jasper*, 79.
48   Reilly, *Wedgwood Jasper*, 78.
49   Reilly, *Wedgwood Jasper*, 78.
50   Reilly, *Wedgwood Jasper*, 79.
51   Reilly, *Wedgwood Jasper*, 90.
52   Reilly, *Wedgwood Jasper*, 90.

CHAPTER 6

1   Details of a trip on the *Turbinia* can be found in Moffett, "Fastest Vessel Afloat."
2   Sir Frederick Bramwell at 1881 meeting of the British Association, as quoted in Smith, "Sir Charles Parsons," 35.
3   Parsons, *Steam Turbine*, 2.
4   Comments from critics in this paragraph are from Parsons, "Application."
5   Leggett, "Spectacle and Witnessing," 288.
6   Parsons quotes in Scaife, *From Galaxies to Turbines*, 364.
7   Dickinson, *Short History*, 146.
8   A superb biography of Parsons is Scaife, *From Galaxies to Turbines*.
9   Scaife, *From Galaxies to Turbines*, 143.
10   See Lyman, "Practical Hero."
11   Parsons, *Steam Turbine*, 8.
12   Quotes from Parsons's wife from Ewing, "Hon. Sir Charles Parson," xxiv.
13   Parsons, *Steam Turbine*, 1.
14   Parsons, *Steam Turbine*, 2.
15   Gladstone, "Henry Victor Regnault," 239.
16   Marsden, "Ranking Rankine," 435.
17   Parsons, *Steam Turbine*, 2.
18   An excellent modern-day description of Parsons's turbine is Scaife, "Parsons Steam Turbine."
19   Parsons, *Steam Turbine*, 1.

20   Parsons, *Steam Turbine*, 2.
21   Wilson, "Global Review at Spithead."
22   "After the Spithead Review."
23   "After the Spithead Review."
24   "The Turbinia."

CHAPTER 7

1   An excellent introduction to the engineering of buildings is Salvadori, *Why Buildings Stand Up.*
2   Ore, "Pascal," 409–10.
3   Boethius, *Consolation of Philosophy*, 32.
4   Smoryński, *Mathematical Problems*, 180.
5   See Singh, *Fermat's Last Theorem.*
6   Kramer, *Nature and Growth*, 293.
7   David, *Games, Gods and Gambling*, 230.
8   The wager is described in Pascal, *Pensées*, Section 418. All quotes are from this section.
9   All quotes from Tippett in this paragraph from Tippett, "Some Applications."
10   Gould, "Smoking Gun of Eugenics."
11   For a general history of eugenics, see Kevles, *In the Name of Eugenics.*
12   The reader interested in assessing Fisher's thoughts on eugenics should consult Bennett, *Natural Selection*; Bodmer et al., "Outstanding Scientist"; Louçã, "Emancipation Through Interaction;" and MacKenzie, *Statistics in Britain.*
13   Quoted in Louçã, "Emancipation Through Interaction," 677.
14   Gumbel, *Statistical Theory*, 56.
15   Details of Thom's life are culled from Collins, "History of Agronomy"; "The Society's Awards"; Marshall and Thom, *Wind Loads*; and Iowa State University, "Memorial Resolutions."
16   "45 Beacon," 712.
17   Thom, "Frequency of Maximum Winds."
18   Coles and Powell, "Bayesian Methods," 119.
19   Cumming, "Risk Assessment," 1.

CHAPTER 8

1   Pope, *Evolution*, 80.
2   Jehl, *Menlo Park Reminiscences*, 2:710.
3   Quotes from Maxim in this paragraph from Maxim, *My Life*, 132.

4   Charles L. Clarke as quoted in Jehl, *Menlo Park Reminiscences*, 2:867.

5   The best starting point to explore Edison's development of his incandescent bulb is Friedel and Israel, *Edison's Electric Light*.

6   Friedel and Israel, *Edison's Electric Light*, 130.

7   "Maxim's Electric Light."

8   Pope, *Evolution*, 81.

9   Jehl, *Menlo Park Reminiscences*, 2:706.

10   Flint, *Memories of an Active Life*, 289.

11   Fouché, *Black Inventors*, 94.

12   As quoted in Cornish, *Machine Guns*, 48.

13   Maxim, improvements in devices for advertising purposes, GB Patent 190702482A.

14   Maxim, *My Life*, 140.

15   Fouché, *Black Inventors*, 91.

16   Fouché, *Black Inventors*, 87.

17   Latimer, process of manufacturing carbons, US Patent 252,386.

18   Nichols and Latimer, electric lamp, US Patent 247,097.

19   "Maxim System," 620.

20   Swinburne, "Incandescent Electric Lamps," 160.

21   Miller, *Yankee Scientist*, 4.

22   Coolidge, "Ductile Tungsten," 961.

23   Miller, *Yankee Scientist*, 55.

24   Coolidge, "Ductile Tungsten," 961.

25   Miller, *Yankee Scientist*, 72.

26   Coolidge, tungsten and method of making the same for use as filaments of incandescent electric lamps and for other purposes, US Patent 1,082,988.

27   Hutchinson, *Big Ideas*, 209.

28   See Johnson, *Shuji Nakamura*.

29   For the full story, see Nakamura, Pearton, and Fasol, *Blue Laser Diode*.

30   Rolston et al., "Rapid Open-Air Fabrication," 2675.

CHAPTER 9

1   Baxter, *Scientists Against Time*, 142.

2   Bowen, *Radar Days*, 155.

3   This fascinating system is well described in Neale, "First Operational Radar."

4   Allan D. White (Raytheon engineer), in conversation with the author, June 16, 1998.

5   Gorn, "Micro Wave Cooking."

6   Murray, "Percy Spencer," 115.

7   Charles Francis Adams, interview, *Invention*, Discovery Channel, February 21 and 22, 1997.

8   This and other quotes in this paragraph from Spencer, "P. L. Spencer."

9   White, conversation.

10   This was reported in Murray, "Percy Spencer."

11   "Rise of Raytheon," 188.

12   Marcum and Kinn, "Heating with Microwaves," 85.

13   Marcum and Kinn, "Heating with Microwaves," 85.

14   Spencer, method of treating foodstuff, US Patent 2,495,429.

15   Wilbur L. Pritchard (Raytheon engineer), in conversation with the author, June 30, 1998.

16   Osepchuk, "History of Microwave Heating," 1,202.

17   Bock's notebooks are in the Raytheon Archives; quotes below are from these notebooks.

18   Raytheon advertisement in *Physical Therapy Review* 31, no. 6 (1951): 258.

19   Richard Ironfield (Raytheon engineer), in conversation with the author, June 17, 1998.

20   For accessible overviews of the problems with evolutionary trees and the importance of cladistics, see Gould, "Ladders and Cones"; and Gould, "Redrafting." Most biologists today seem to regard cladistics as arcane in the age of DNA; for a spirited defense, see Brower, "Fifty Shades of Cladism."

21   This example is drawn from Cowan, *More Work for Mother*.

22   Hughes, *American Genesis*, 4.

23   Rushkoff, *Program or Be Programed*.

24   Mumford, *Technics and Civilization*, 282.

25   Richard Sennett's *The Craftsman* is a book-length rejoinder to Arendt, his mentor. In it, he fairly outlines her approach, but the reader can find the fullest description in Arendt, *Human Condition*.

26   Perhaps the most prominent current technological utopian is Kurzweil, *The Singularity Is Near*, although this type of thought can be traced to the nineteenth century; see Segal, *Technological Utopianism*, which is a fascinating study of twenty-five utopian novels written between 1888 and 1933.

27   Simon, *Ultimate Resource 2,* 589.

AFTERWORD

1   Bush, *Science—The Endless Frontier*. Bush's notions of science, engineering, and technology were broader than the linear model extracted by

universities from his manifesto; see Leyden and Menter, "Legacy," for a nuanced take.

2   The discovery of the mold *Penicillium notatum* by Alexander Fleming predates World War II by a decade or more, but it was never mass manufactured, although several pharmaceutical firms tried and abandoned production. In 1941, the U.S. government initiated a program to manufacture penicillin, and by January 1945, they were producing four million doses a month. See Quinn, "Rethinking Antibiotic Research."

3   For a fascinating study of the idea of a "lone" inventor for patent law, see Lemley, "Myth of the Sole Inventor." For a review of historians' criticisms of popular accounts of invention and technology, see Bray, "How Blind Is Love?"; Gascoigne, "'Getting a Fix'"; Knell, "Palatable Myth of William Smith"; and Miller, "The 'Sobel Effect.'"

4   Bronowoski, *Ascent of Man*, 15.

5   Petroski, *To Engineer Is Human*.

APPENDIX: THEMES

1   *Oxford English Dictionary*, s.v. "conservative," accessed February 8, 2022, https://www.oed.com/view/Entry/39569.

ACKNOWLEDGMENTS

This book draws on insights from a lifetime of teaching and thinking about engineering, which means that details of the intellectual debts and assistance are lost in memory—thousands of students, hundreds of colleagues, many collaborators, and innumerable librarians. I can, though, enumerate the more recent debts owed to those who helped develop and prepare this book.

My most significant and direct debt is to Chris Francis. While an editor at Sourcebooks, he acquired this book and then, after leaving Sourcebooks, continued as a freelance editor. Any clarity of theme is from his urging, at every juncture, "More theme! more theme!" In addition to helping erect this overarching superstructure, he detected fuzzy thinking, unclear explanations, and confusing sentences. His work was fine-tuned by Jenna Jankowski at Sourcebooks, whose careful eye uncovered many infelicities.

I also owe a debt of gratitude to an anonymous group of

editorial decision makers at the highest levels of Sourcebooks. With Chris's help, I spent months preparing a book proposal—a forty-page or so document—to pitch a science and technology subject. When Chris brought it forth, they rejected it! The message from those decision makers was simple: "He has a better book in him." While at first angry at this rejection, I did, working with Chris, pitch *this* book—indeed a much better book!—so I thank the editorial team at Sourcebooks for their insightful rejection.

And last, yet another thanks to an anonymous group of people: the groundskeepers at nearby Japan House, whose work kept the book alive when the COVID-19 pandemic halted my work on its manuscript. I sheltered in place for months with my wife and two young sons. My wife was essential personnel and worked throughout, so my focus was on childcare. My sons and I toiled away the days at Japan House looking for frogs in the ponds, climbing trees, and playing hide-and-seek as we waited for vaccines to be developed and deployed. For many months, little writing was done, but as the boys frolicked for hours in the vast grounds, I could often snatch a few moments to think about this book. That was just enough to keep the thread from breaking, although it was stretched to the limit so often that I thought the book might vanish from my mind, never to be written.

ABOUT THE AUTHOR

 Bill Hammack hosts the *engineerguyvideo* YouTube channel, which has over a million subscribers. *Make* magazine said of Bill's video work that he was a "brilliant science-and-technology documentarian" whose "videos should be held up as models of how to present complex technical information visually." *Wired* called them "dazzling." He teaches engineering at the University of Illinois at Urbana-Champaign, where he focuses on educating the public about engineering and science. Bill's work has been recognized by a broad range of engineering and scientific societies, and he has received the Hoover Medal, awarded jointly by five engineering societies, the Public Service Award from

the National Science Board, and the Ralph Coats Roe Medal from the American Society of Mechanical Engineers. He has also won the trifecta of top science and engineering journalism awards: the National Association of Science Writer's Science in Society Award, the American Chemical Society's Grady-Stack Medal, and the Science Writing Award from the American Institute of Physics.